A

COMPLETE TREATISE

ON

ARTIFICIAL FISH-BREEDING:

INCLUDING THE REPORTS ON THE SUBJECT MADE TO THE FRENCH ACADEMY AND THE FRENCH GOVERNMENT; AND PARTICULARS OF THE DISCOVERY AS PURSUED IN ENGLAND.

TRANSLATED AND EDITED BY

W. H. FRY.

ILLUSTRATED WITH ENGRAVINGS.

NEW YORK:
D. APPLETON AND COMPANY,
346 & 348 BROADWAY.
1866.

The following is a table of some French measures used in the course of this work:

A Metre	is equal to about $3\frac{1}{4}$ feet.
A *Kilomètre*	" " $1093\frac{1}{2}$ yards.
A *Decimètre*	nearly 4 inches.
A *Centimètre*	about one-third of an inch.
A *Millimètre*	about one twenty-fifth of an inch.
A *Hectare*	about $2\frac{1}{2}$ acres.

The *mètre* is the base of all French measure, and is a fixed or scientific means of determining length adopted by the French Institute and Government: it is the ten-millionth part of the arc of a meridian between the pole and the equator.

PREFACE.

THIS work has been prepared at the suggestion of a number of my scientific friends. Such a treatise is needed, as various agricultural societies* of the Union, and

* ARTIFICIAL FISH.—It is not often that the proceedings of the Farmers' Club, at the American Institute, supply a theme worthy of much consideration. A practical farmer occasionally gets among the judges, doctors, professors, &c., and relieves their forlorn incoherences with a few sensible suggestions and useful information; otherwise the discussions of the fancy city-agriculturists who assemble in the Institute rooms are very incongruous affairs—grateful, no doubt, to the obfuscated vanity of a few bewildered fogies, who can there make speeches on subjects of which they know nothing, or read dull translations from the French or Dutch, on other subjects, respecting which nobody cares anything. But some new blood has evidently been infused into the Club, and its proceedings are becoming worth the space allotted to them by the newspaper reporters. At a recent meeting, the subject of the artificial propagation of fish was introduced—and a more important one could scarcely engage their attention.

The disciples of Isaak Walton have long deplored the gradual extinction of trout in almost every stream in which they formerly abounded, within a day or two's ride of the city. The rapacity with which they are pursued and caught, when too small for any likewise the Legislature of this State,* have shown themselves alive to the importance of the new discovery, and hence indicated the necessity of a manual on it.

purpose—too paltry even to "send to market"—argues a coarse spirit of destructiveness that cannot exist in the true lover of angling. We know parts of Connecticut famous for trout and game not many years ago, where neither a trout nor woodcock can now be found, and even the quail is a traditionary bird; yet there are fine streams, and plenty of woodland. The unchecked lust for shillings has not left a fish or bird in whole counties. So, too, on the south side of Long Island, once esteemed among the best trouting localities in this State, where mischievous boys, and vulgar men, have been allowed to destroy them, until now a trout can scarcely be found. Nor are these worse than some of our city "sportsmen!" whose highest idea of sport is wanton destruction. We heard one boast last summer of having killed twelve hundred trout in two days at Catskill! Of course they were nearly all young fish, the largest probably three inches long. A very few were brought home (putrid when they arrived), and the remainder were left to perish on the banks of the stream. We wonder the people of that region did not prevent such vulgar slaughter. The man ought to be prohibited from all "sport" but catching bull-frogs for ever after.

Mr. Pell, the celebrated horticulturist of Ulster county, who has also turned his attention to pisciculture, predicts the speedy and total annihilation of shad in our rivers, unless something is done by the Legislature to preserve them. He avers that whereas it used to be a common thing to draw 1600 shad at a haul, the fisherman now gets sometimes one or two fish in his net. These are facts coming directly home to our breakfast and dinner tables, and certainly are legitimate subjects for legislative interference.—*From the Sunday Times, March* 19, 1854.

* Preservation of Fish.—A bill has been introduced into the New York Senate for the preservation of fish in the waters of this State. The first section provides that "all persons in the State of New York who obtain a livelihood by the capture of fish, shall, toward

All so far known that is really important in regard to it, and has been brought to light by the commissioners appointed by the French Government to investigate it, as well as by the experiments of private persons in France, is contained in this volume. To this is added a statement of English pisciculture. The entire history and practical details of artificial fish-breeding are, therefore, to be found in these pages. The value of the discovery, and the expediency of turning it to account, will speak for themselves. According to the authorities I have cited, it is a subject equally interesting to the farmer, the economist, and the statesman, and will prove a source of immeasurably great wealth if properly pursued.

It seems that a discovery of the highest importance, of a mode of actually creating fish in illimitable numbers, was made in Germany nearly a century ago; but so much occupied were the people of Europe in the art and science of cutting one another's

the close of the fishing season, impregnate the spawn taken from at least two dozen female fish, with the milt taken from the same number of male fish, and plant the same upon their fishing ground, in presence of the justice of the peace of the district, or some person appointed by him." A fine of $50 is provided for a violation of the act. Suits may be brought by the superintendents of the poor, and the penalty to go to the support of the poor. If brought by other persons, one half the sum to go to the support of the poor, the other to the person bringing the suit.—*From the Sunday Atlas, March* 26, 1854.

throats, that it was lost sight of, and was recently re-discovered in a more valuable form, by two poor illiterate French fishermen, and practically demonstrated by them more than ten years since, in their quiet and secluded neighborhood. Discoveries, however, to be known in France at all, must, owing to the system of centralization existing in that country, nave Parisian and governmental sanctions. So it is not surprising that this one of artificial fish-culture, remained unnoticed and unknown till 1849, when, having chanced to come to the knowledge of Dr. Haxo, a scientific man residing in the same department as the two fishermen, it was by him communicated to the Academy of Sciences at Paris, in a paper which caused a great sensation in that learned body. Immediately a commission, consisting of three of its distinguished members, Messrs. Milne-Edwards, Duméril, and Valenciennes, was appointed to investigate the discovery, and make a report. Their investigations, as well as those of the celebrated naturalist, Mons. Coste, brought out the fact that the discovery was only an improvement upon an old one which was made and forgotten nearly a century ago. The merit of the two fishermen, Géhin and Remy of Vosges, was, however, undeniable, and admitted by the commission; and there appears to be no doubt that theirs was a veritable discovery, or re-discovery, and that but for them the method of artificial fish-culture would have remained to this day as unknown as it was twenty years ago.

The proceedings of this learned body, the highest scientific authority of France, and perhaps of the world, brought the matter to the attention of the French government, and resulted in the appointment, by the Minister of Agriculture and Commerce, of M. Milne-Edwards, a member of the Institute, to make an investigation, and report to the Government. This report is the subject of a volume of criticism by Dr. Haxo, who censures Milne-Edwards and the other academicians for their unfairness to the two fishermen. Be that as it may, the Government rewarded the fishermen with places and pensions: and upon the recommendation of the Report that theirs was the best method, but from want of means they had not been able to carry it out on any large scale, the Government determined to take the matter in hand. With this view it made a first appropriation of 30,000 francs, and appointed Messrs. Berthot and Detzem, the engineers of the Rhine and Rhone canal, to erect at Huningen a government-establishment for artificial fish-culture, and to superintend its workings. This establishment went into operation in 1852, and according to the report of M. Heurtrier, Director-General of Agriculture and Commerce, made to the Minister of the Interior, its superintendents, Messrs. Berthot and Detzem, had, in the first six months, artificially fecundated 3,302,000 eggs, and produced 1,683,200 living fish. Of these, according to M. Coste, 600,000 were of the valuable species trout and salmon.

These results achieved, under Government sanction, brought public attention to the subject in France and in Great Britain. Seven works, by as many different authors, were published upon the subject: three in France by Coste, Godenier and Haxo; and four in England by Shaw, Boccius, and two anonymous writers. Agriculturists and general readers on this side of the ocean began to wish for light, and among others the editor of this volume. But none was to be had, for strange as it may appear, yet true, and proving how slowly the knowledge of great truths sometimes travels, not a copy of any one of these publications was to be found at any bookseller's in New York, even a year after their publication in France and England. The editor consequently caused them to be imported; and, translated by him, in this single volume will be found a selection embracing all that is valuable in their contents.

This book contains a translation entire of the work of M. Coste, the fullest which has appeared, giving a history of the discovery and of all that has been effected by it; likewise a complete translation of the pamphlet of M. Godenier, describing particularly the practical methods of the two fishermen, Géhin and Remy, for whom he claims the merit of the discovery; a translation is also given from Dr. Haxo's work, of M. Milne-Edwards' Report to the French government upon the whole subject; and to these is added a series of papers on artificial salmon breeding published in *Bell's Life in London*, during January and February of this year, and containing some valuable information on that particular branch of the subject.

The discovery of artificial fish-culture, in a word, claims to show how, at little care and little cost, barren or impoverished streams may be stocked to an unlimited extent with the rarest and most valuable breeds of fish, from eggs artificially procured, impregnated and hatched.

W. H. F.

NEW YORK, July, 1854.

Plate I.

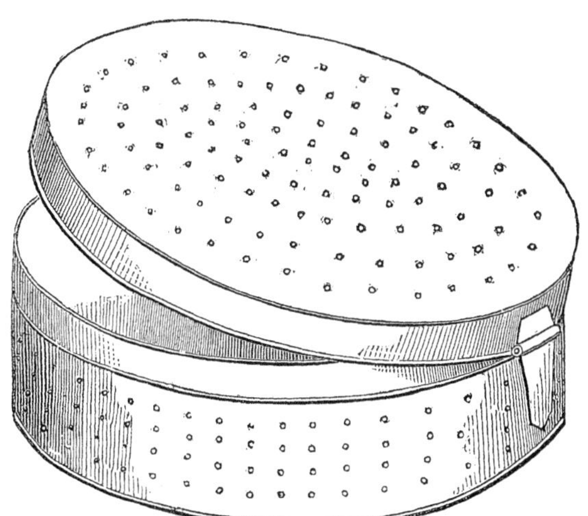

HATCHING-BOX OF GÉHIN AND REMY.

(*See page* 14.)

Plate II.

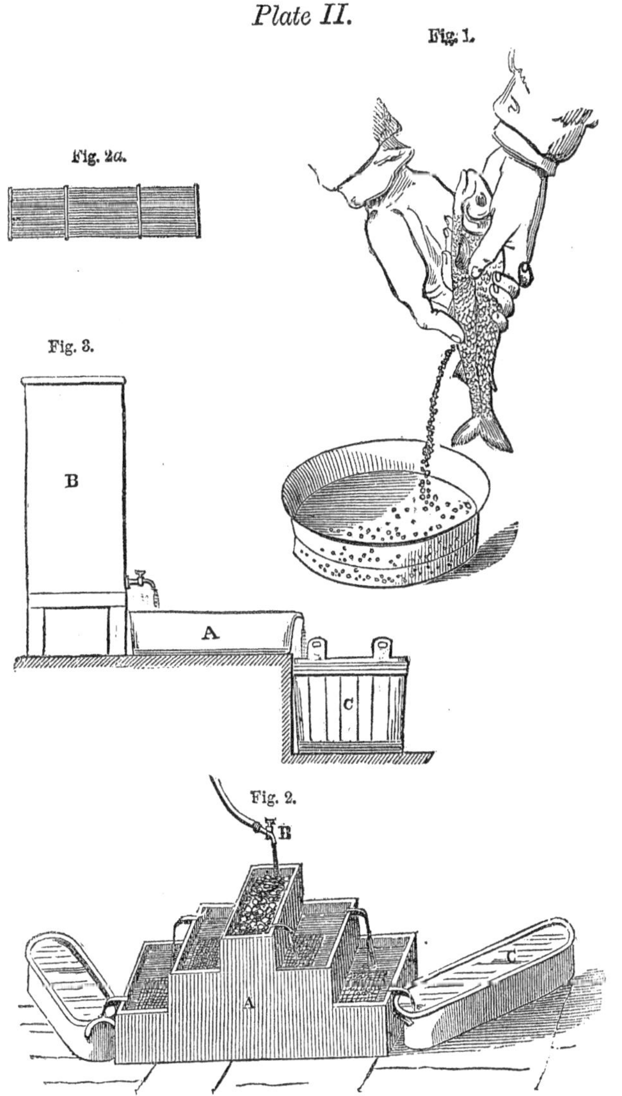

EXPLANATION OF PLATE II.

FIG. 1.—Figure intended to give an idea of the manner of extracting the eggs from the living fish.

FIG. 2.—Working-apparatus for hatching: *a*, parallel canals composing it; *b*, stop-cock by means of which the stream of water can be regulated; *c*, long earthen tubs adapted to this apparatus for receiving the fish just hatched.

FIG. 2*a*.—Front view of hurdles upon which the eggs are placed in the hatching apparatus.

FIG. 3.—Another kind of hatching apparatus, consisting of a canal (*a*) of wood or earthenware, of an ordinary fountain (*b*) and a tub to receive the running water, (*c*.)

Plate II., continued.

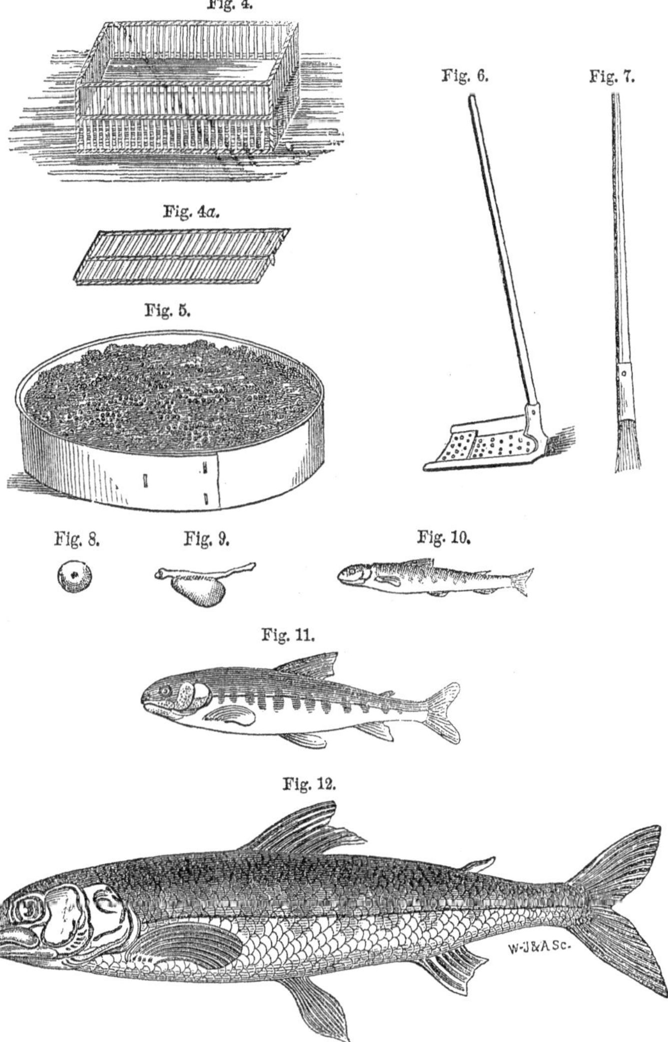

EXPLANATION OF PLATE II., Continued.

Fig. 4.—Floating apparatus for hatching in running streams, consisting of a wicker-box or basket, in which are fitted hurdles, as in the other apparatus.

Fig. 4*a*.—Hurdles for the floating apparatus.

Fig. 5.—Box for transporting the eggs.

Fig. 6.—Small perforated leaden shovel for lifting from the hurdles the fish just hatched, to transfer to the tubs or running streams.

Fig. 7.—Brush of badger-hair for the purpose of cleaning the eggs of the sediment which settles upon them.

Fig. 8.—A salmon's egg developing.

Fig. 9.—A salmon just hatched; *a*, its umbilical bladder.

Fig. 10.—A young salmon having been fed from its umbilical bladder.

Fig. 11.—Young salmon, aged three months, reared at the College of France.

Fig. 12.—A young salmon, aged six months, reared at the same place.

TREATISE

ON

THE ARTIFICIAL FECUNDATION AND INCUBATION

OF

THE EGGS OF FISH,

AND

THE REARING OF THE YOUNG FISH, ACCORDING TO THE PROCESSES OF MESSRS. GEHIN AND REMY, FISHERMEN OF VOSGES, AND MEMBERS OF THE ACADEMY OF SCIENCES OF PARIS.

PREPARED FROM THE FACTS FURNISHED

BY

M. GEHIN

TO

THE EDITOR, C. E. P. GODENIER, FISHERMAN.

M. GEHIN IN THE DEPARTMENT OF ISERE.

MR. GEHIN, one of the authors of the great discovery of the artificial production, fecundation, incubation and hatching of the eggs of fish, was commissioned by the government, at the instance of Mr. Adolph Perier, to make a tour through the Department of Isère, in the month of November, 1851, for the purpose of instructing others in the processes by which he accomplished such extraordinary results. He travelled through many of the communes, stopping at Allevard, Pontcharra, Sassenage, Veurey, Vizille, D'Oisans, Rives, Pont-en-Royans, Paladra, Lemps, St. Geoire, Arandon, Buisse and Grenoble, and every where experimenting publicly. He enlisted the services in some places of Messrs. Janon and Rafin, fishermen of Veurey, who were engaged in collecting the spawn from which supplies could be furnished as required.

He also made arrangements with Mr. Millon, fish dealer at Charavines, by which he would be able to furnish the eggs of *l'ombre chevalier ;* a very rare species of trout, which, in France, is only found in Lake Paladru, and indeed only in a certain part of that lake. This is the most beautiful of French fishes, and by epicures is considered more delicate than any other taken in fresh water. It spawns in December. Mr. Perel, fisherman, No. 3, Neuve-des-Pénitents street, Grenoble, can supply the spawn of this trout.

In the course of his journey, Mr. Gehin prepared more than two hundred boxes of eggs and deposited them in well selected places to effect the artificial hatching.

NATURAL PRODUCTION AND HATCHING.

We have witnessed the operations of Mr. Gehin, at Sassenage. But before describing them, let us say a word on the causes which led Messrs. Gehin and Remy to discover their method. Mr. Gehin has detailed them to us, as follows :—

Messrs. Gehin and Remy, of the commune of Bresse, in the department of Vosges, were fishermen, living by their calling. Every year at spawning time, they felt regret at the vast destruction of eggs contained in the female fish they took, and naturally their thoughts turned towards the discovery, if possible, of some mode of preventing the evil.

As long ago as 1841, they commenced to observe carefully the habits of the trout, and in the month of November of that year, during a full moon, they passed night and day on the bank of a river, never for an instant losing sight of these fish, and watching most intently all their preparations for laying and preserving their eggs.

The results of their observations were these :—

The trout come together in a shoal, and choose a current with a gravelly bottom as the best place to lay their eggs. They dig in it a round hole, sometimes of the depth of seven inches by three feet in diameter ; they place in the middle of this space, parallel with the current, a line of stones, the size of which varies with the size of the fish.

The female then passes over this line of stones, gliding over, rubbing against or resting upon them. This she does again and again, some twenty or thirty times, till her eggs are all laid in the crevices of the gravel.

When the female has done this, the male, in

the same manner, by passing over and pressing upon the gravel, emits the milt, or soft roe, which covers and fecundates the eggs ; then with tail, fins, head, and belly he works away till he manages to cover the eggs with gravel.

Now a second female commences, and in the same manner lays her eggs in a parallel line with and against the first row. When the fecundation is complete, which generally happens in about fifteen days, according to the number of fish, all unite in heaping up stones and gravel in mounds upon the eggs, in a manner resembling the great ant hills that may be found near by.

Mr. Gehin believes that their mason-work is, in a manner, cemented by a slimy secretion, with which they cover the stones, while incessantly rubbing over and pressing against them in heaping them up ; for he found it difficult to destroy the mounds so formed by scratching apart the material with his fingers.

The eggs remain in this way for a month or two, while the process of incubation goes on ; at the end of a time which Mr. Gehin could not precisely determine, the little fish appear about the size of pins, come out of their cell between the interstices of the gravel, and seek in the tranquil waters, near the shores, a place of safety.

Having thus discovered nature's secrets, it remained to discover a mode of rendering them practically useful, and not until after many failures did Gehin and Remy hit upon a sure process, incontestably superior even to that of nature herself. This may be deemed too bold an assertion, but a moment's reflection will prove its truth.

Does not man sow his entire field with the single sort of grain he wishes to cultivate? Where in its natural state will we find a field producing one only kind of grain? Do we ever find in a state of nature, and within equally circumscribed limits, the same number of any single species of animal, as in our stables, sheepfolds, or barnyards? What, then, is there to hinder us from stocking bountifully our streams with fish, by aiding the process of hatching, and protecting the young from destruction by their innumerable enemies?

How these objects are attained by Mr. Gehin we shall now proceed to show, regretting to add that his partner, Mr. Remy, is no longer able to co-operate with him, owing to a disease—the result of exposure during the experiments—which bids defiance to doctors' skill and incapacitates him from labor.

ARTIFICIAL FECUNDATION.

We will now proceed to explain the processes of Mr. Gehin, as exhibited at Sassenage. His experiments there were made with trout of the weight of 250 to 300 grammes, preserved alive in a reservoir.

Mr. Gehin takes a female trout when she is ready to spawn, holding her by the back with his left hand, he prevents her violent struggles by pressing her head and body against him, and with his other hand strokes her belly, till, in a few moments, she becomes quieted. All animals are sensible to these caresses, or similar ones made on their backs, and take them willingly ; witness the cat and dog, which by purring, whining, rubbing against us, or licking our hands seek to obtain them.

When the trout is thus magnetized, or put to sleep, as Mr. Gehin calls it, he inclines it over a vase, which he has prepared to receive its eggs by putting in it about a quart of water ; in order to make sure of the fish remaining quiet, another person, if necessary, holds its tail, then Mr. Gehin with the thumb and forefinger of his right hand presses lightly the belly from top to bottom. This must be carefully and gently done, as one would press from root to extremity a finger cut at the end, to extract the blood and prevent its further flow ; or as one would milk a cow, but by no means with so much force as that operation requires, for if the proper time has been chosen the eggs will be pressed out by a very gentle effort, and if more is required, it will prove that the fish has not gone her full time for spawning, and the eggs thus obtained cannot in that case be fecundated.

Passing the finger and thumb in this manner over the fish's belly, the eggs at each pass will spirt out, like a little liquid stream, falling into the vase.

When by a number of these passes the eggs are

all pressed out, a male fish is taken and operated upon in the same manner ; the milt thus expressed from the male falling into the vase and upon the eggs gives the water a white hue. The male fish, like the female, must be subjected to a number of gentle passes to obtain the result. When this is done, the contents of the vase must be stirred about with the hand, or what is still better, with the tail of the male fish still wet with the milt that has flowed over it,—an operation resembling that made by the fish in its natural state.

After a very short period the water must be carefully poured off and a like quantity of fresh water poured on the eggs.

Before the mixture of the milt and the water covering the trout's eggs, their color is a pale orange and transparent. After the mixture the eggs that have been fecundated assume a brownish hue, and a black speck, of the diameter of two millimetres, appears in the centre of each.

After this the water must be changed once or twice more.

When the fecundation is complete some of the eggs will appear white. These are the unfecundated ones being sterile and dead, and if allowed to remain will, like all other lifeless things, become putrid and corrupt the rest ; they must therefore be carefully removed.

ARTIFICIAL INCUBATION.

We have seen that in artificial fecundation nature has been imitated, where the fish voids its eggs by passing over and pressing its belly against smooth stones. This natural method being imitated by the hand of the operator all the eggs are evacuated without the loss of a single one.

In artificial incubation Mr. Gehin has in like manner exactly imitated nature in the processes he employs. He takes a round box in the form of a warming-pan, except that the bottom rises a little in the inside in order to make it remain more firm in the position in which it may be placed. It is made of zinc to prevent rust ; its size—twenty centimetres in diameter and seven centimetres in depth; the lid four centimetres in height, on hinges, with a catch. The box is pierced on every side, with two thousand holes, so that the water can freely flow through it over the gravel. These holes are a millimetre in diameter and should be very carefully and smoothly made, with a punch, in such a manner as not to wound the little fish attempting to escape through them.

The bottom of the box is covered with a bed of fine gravel, and on this are placed the fecundated eggs. Each box should contain one brood of eggs. The box is then closed, a hole is dug for it in the gravelly bottom of a running stream of fresh water in which it is placed and gravel is strewed over it.

It is necessary to take all these precautions in order that the water flowing through the gravel may be purged of its impurities before it enters the box, and not deposit mud and slime upon the eggs and retard or prevent altogether the hatching; as Gehin and Remy observed was the case when such precautionary measures were omitted.

The box so placed is left for a month or two. Mr. Gehen could not determine the exact time of the process of incubation, as it varied with the quality of the water. This question is now occupying the attention of scientific men, and from their researches we shall have the exact results.

In place of using a box, a hole may be dug in the gravel, and the eggs may be deposited in it and covered over with pebbles in the manner practised by the fish. But the progress of the incubation cannot then be as accurately watched as with a box, which can be opened and closed at pleasure.

Mr. Gehin observed these phenomena of the hatching: the tail comes first from the egg, and the pieces of the fine skin or shell torn by it, form the two hinder fins. The head next appears at the other end, and the torn shell there forms the forward fins. The lower part of the egg forms the belly, and the upper part next is broken, and the back appears. The shell or skin which enveloped the embryo is not detached from the newly-born fish, but becomes a part of and is absorbed by it.

PRESERVATION OF THE YOUNG FISH.

As incubation goes on and hatching time approaches, in order to determine when it will take place and not to their injury retain the little fish too long prisoners, the boxes should be frequently inspected. When incubation is finished and the little fish begin to move, they must still be kept inclosed from eight to fifteen days, according as their numbers are small or great; then they may be set at liberty in the quietest part of the stream, care being taken that the quality of the water is the same as that which has flowed through the boxes, for a change to water of more or less freshness or clearness will have a sensible influence on their frail existence.

A wide field for experiment is here open to any who choose to enter upon it: for example, it would be very interesting to ascertain the results of different modes of treatment by placing similarly, either confined in large boxes or at liberty in running brooks, equal numbers of young fish, and supplying one set with thickened blood and other food, and leaving the other set to find such food as nature affords, to note the relative increase. Experiments, too, might be made with boxes having larger holes than those before described—large enough for the escape of the young fish as soon after hatching as they might seek their liberty; which mode might do away with the necessity of watching the process of hatching, with the view to opening the boxes soon after.

ON THE PRESERVATION AND TRANSPORTATION OF THE EGGS.

In some sections of country, when many crops have been gathered from the same plot of ground, it becomes in a degree exhausted, and, whether to improve it for the same or prepare it for another kind of crop, it is usual to overflow it, converting it into a fish pond and stocking it with carps and tench. After three or four years the water is drawn off, the fish taken and sold, and the land cultivated. A few years having elapsed it is again overflowed, but need not again be stocked with fish, as they now appear, as it were, spontaneously. We have the evidence of numerous observers as to this fact, and among others the inhabitants of a commune near Grenoble. They state that there is a little lake on the side of a hill in the commune of Jarrie, not far distant from their town, which is well stocked with carps and tench. Sometimes the lake dries up and the ground is then cultivated with hemp and yields abundantly; and again it is filled with water, and carps and tench appear in great quantities.

The river Drac has a wide bed, and the current at times makes new channels or retakes old ones which have dried up; in the pools of these old channels soon appear quantities of trout whose size shows their age to date from the period of the water's return. This fact is vouched for by the fishermen of that river. The inference is clear, that the fecundated eggs have been left in the bed of the channel at the time it dried up, and so remained until the returning waters have united all the necessary elements of incubation, which has then taken place, and soon after the eggs have been hatched.

Founded on these facts, Mr. Gehin adopts the following mode for preserving and transporting fecundated eggs.

He takes one of the boxes we have already described, and covers the bottom with a bed of fine sand; he covers this with a layer of gravel or pebbles, varying in size from a pea to a hazelnut, and upon the gravel he spreads a layer of fecundated eggs. These again are covered with sand, then another layer of gravel and then one of eggs, and so on till the box is filled. He takes great care beforehand to wash the sand and gravel so as to free it entirely from every particle of mud or slime.

When the box is full he dips it in water so that its contents shall be thereby more closely packed together; and being thereafter exposed to the air it can be sent anywhere without altering the condition of the eggs.

On the receipt of the box its contents are taken out and placed in five or six other boxes—each containing the spawn of one female—and these are placed in the gravelly bed of a stream according to the directions before given as to artificial incubation.

If any of the eggs have a different appearance from that before described as fitting them for incubation they must be thrown away, and the cleaner the eggs are the better will be the success of the experiment. The sterile and spoiled eggs become white and opaque like the white of a hen's egg boiled, and if broken yield a milky fluid.

Mr. Gehin tried the experiment of drying in the sun eggs spread on paper, and then placing them in a hatching box. With some he succeeded. Now that Gehin and Remy have opened the way, we may look for the results of various experiments that will doubtless be made by others, in the modes of preserving eggs, crossing the breeds of different species of fish, and kindred matters.

UTILIZATION OF FISHING AT TIMES PROHIBITED BY LAW.

Trout-fishing is prohibited during spawning time, that is during November and December, because the fish coming then together in shoals can so easily be taken and the work of reproduction prevented. But fishermen, stimulated by large profits and lessened toil, brave the law and destroy in the germ millions on millions of fish.

To these fishermen Mr. Gehin gives this advice: they should take with them a little pot made of zinc;

say five inches in diameter and six inches deep. It should be filled one third full of water, and a piece of wood the size of the pot may cover the water to prevent its splashing out as the pot is carried from one place to another by the fishermen while at their work. When they take a female fish they should put her eggs in the pot and in like manner put into it the milt from the male fish. Mr. Gehin thinks it unimportant whether the eggs or the milt be first placed in the pot; that will depend on whether the female or male is first caught. When they have done fishing they should dig a hole in the bed of the current near where they have been fishing, and cover the bottom with fine gravel. Into it they should pour the eggs, which, when fecundated, will drop to the bottom among the gravel like little shot. The hole should then be filled up with little stones and covered with gravel, and incubation will go on and in due time the eggs will be hatched.

By these means the fishermen, while reaping all the advantage of their illegal trade, will not be in fact doing any injury to the process of reproduction.

GENERAL REMARKS.

Mr. Gehin has observed that fish are with difficulty acclimated on being taken from one stream to another ; often they die ; almost always they become sterile. But eggs so transported are easily incubated and hatched, and produce fine fish.

He has also remarked that except in a running stream, other fish than carp and tench become barren.

When fish are caught for the purpose of taking from them the eggs and are found to be in an unfit state to be operated upon, from not having gone their full time, they should be kept in a reservoir till the proper time, and then relieved of their eggs and set at liberty again. Otherwise they will die in the reservoir; as they will not while in that manner kept prisoners spawn naturally, but will retain their eggs and perish.

When eggs are preserved in water, it must be frequently renewed or aquatic plants be placed in it. These plants preserve the eggs in unchanged condition.

We have weighed one hundred unfecundated eggs of small trout, after having preserved them in water for two days, and dried them on linen for five minutes, and found their weight to be four gram. Twenty-six eggs placed in a straight line made one hundred millimetres, which gives a diameter of a little less than four millimetres for each egg.

RESULTS WHICH MAY BE HOPED FROM THE DISCOVERIES OF GEHIN AND REMY.

Man reigns on the earth supreme: he bends the sun to his use for the plants he requires; domestic animals submit to his will and produce at his pleasure; he commands the waters to transport him with fearful speed, and soon perhaps the air will be conquered. All nature seems to obey his laws. The fish alone have escaped his dominion, but not his nets.

Now Mr. Gehin has discovered the secret of their reproduction, and placed in our grasp the means to enliven our rivers and watercourses, as we cover our fields with corn, hemp or flax, as we multiply our flocks, our domestic fowls, our silk worms.

The discovery of Gehin and Remy is a great fact for humanity, parallel with the introduction into France of the potato.

Like all other great discoveries, now that we have it, this seems the simplest in the world. How, we ask ourselves, was it possible for any one to eat fish-roe without thinking of the innumerable fish thus destroyed in the germ, and what would that thought lead to but the search for a mode of preventing such wholesale slaughter and the easy discovery of such a mode? Yet it took six thousand years to find the right means to solve readily and practically the difficulty; and we now have the solution in so simple a form that even the children of fishermen practise it as easily as they would tend a flock of sheep.

But the most elementary ideas, which ought to open at once the eyes of the world, are often those which are last taken hold of. They seem, indeed, like the mystical and undecipherable finger-post inscription "*cesticilechemindesanes*"* to the bewildered gaze of the peasant.

Thus, whenever and wherever on this earth three men have been brought together, in what is called the social state—that is, one to command, the second, the office-holder to oversee the execution of commands, and the third, the hewer of wood and drawer of water, to do and to suffer,—poor humanity has revolted, and hence the ever-enduring, universal social war, for a just distribution of nature's gifts. Yet the solution of this problem, which has caused the flow of oceans of blood and tears, is perhaps as simple as that of artificial fecundation; and possibly should society and government devote to it ever so small a part of its vast machinery of subjugation and repression, the mighty social problem would soon be resolved.

For ever will the discovery of Gehin and Remy be a fruitful fact for humanity; one of the grandest discoveries of ancient or modern times; a discovery which we place even above that of Leverrier: for, to be like him, the philosopher, man must live, and they find new and abundant means for existence. And this we aver while signalling Leverrier's discovery as the most brilliant evidence of the grasp and power of the human mind. And even with this acknowledgment, we would still put above his marvellous work

* "This is the road for Asses."—*Tr.*

the discovery of a sure mode to prevent all diseases incident to the potato or the grape ; for potato-roots saved from destruction would assuage much human misery, while the knowledge of the existence of Leverrier's planet lessens not a single pang of a single human creature.

Leverrier has been loaded with honors and crosses, he and his, and he earned these rewards. But what has society done for Gehin and Remy ? It has given them both together the one sixth part of the sum offered to Mr. Gehin by the Spanish ambassador if he would take his invention to Spain. To Mr. Gehin the government has given an appointment to travel, as it may direct him, and spread abroad through France a knowledge of his discovery, and this office has a salary about equal to that of a petty commercial traveller. But, modest as is the salary, it suffices Mr. Gehin, who is distinguished for modesty and sobriety—qualities which add tenfold to any income. All the gold in the world cannot bring to such a man the riches which he finds in the moderation of his desires.

But when we see so many men filling honorable and well-paid functions, loaded with pensions and decorations after a few years' service, as if it were difficult to find men capable of doing such duties, we naturally inquire how it is that the great work of Gehin and Remy has not obtained for them a cross of honor. What social fact of our time can be put in comparison with their discovery?

If the smallest pattern of an academician—one called to his chair for a successful piece of poetry—had made such a discovery, he would by this time have had crosses from every sovereign in Europe ; but these are only two fishermen !

The bestowal of the cross, when it is richly merited, honors the giver ; and, in any case, tends to elevate the receiver. Not only crosses, but statues are due to Gehin and Remy, and posterity, more just than their contemporaries, will do them right.

INJUSTICE TOWARDS THE INVENTORS.

On the announcement of Gehin and Remy's beautiful discovery, the scientific world was of course in commotion. Envoys from the academies of Paris, Holland and Strasburg, came to the inventors to be assured of the reality of their claims. Reports attesting the truth were made to the Academy of Sciences and the government. Mr. Gehin was called to Paris. There he met the most honorable and friendly attentions from his countryman Mr. Buffet, the Minister of Agriculture and Commerce, from whom he received many dinner invitations. He had also the honor of being invited to dine with the President of the Republic. Besides all this, the Academy of Sciences inscribed the names of Messrs. Gehin and Remy among its members.

All these circumstances denoted very clearly that

the process discovered by them is an economical fact of the highest importance. Simple fishermen are not thus treated, who have nothing but their industry to depend on for their livelihood, and no circle of acquaintance beyond their humble position in the social scale. But in the breasts of these two men existed instinctively the love of humanity, that fruitful sentiment which gives birth to great ideas and facts, the product of the heart and not of education.

In the presence of such results the envy of savants soon developed itself, and ere long some were found contesting with the inventors the claim to their discovery. Five or six learned writers were cited, who, in as many memoirs in different languages, produced in the course of a hundred years, had shown that there existed some vague notion as to the practicability of artificial fecundation. And on this basis our two fishermen were accused of plagiarising the original idea—they who in their process had never studied any other than the book of nature.

We are greatly astonished at the injustice of these learned men, whose position should have prevented them from such acts. It is well known that learning does not invent : that is the province of genius. The most beautiful ideas are not due to science ; they are hatched from privileged brains, where they spontaneously appear, without regard to the knowledge of their possessor. Study does not conduct to ideas, it only leads to consequences.

In the present case it seems to us, that, so far

from blazoning to the world what those meritorious writers may have said about artificial fecundation, our savants should rather have been careful to keep their facts out of view ; for if we admit them, they only go to prove that these authors having communicated to the academies their knowledge of artificial fecundation, those learned bodies could not put together three simple ideas : *fecundation, artificial hatching, and propagation,* and draw from them, for the interests of humanity, the natural results which must strike the sense of any intelligent person ; and that it needed two simple fishermen to vivify their ideas and spread them before the world.

How, then, can this embryo thrown a hundred years ago into the domain of science, and which up to this day has never been able to pierce the dust that envelopes it, be compared with the living idea of Gehin and Remy ? an idea which from the beginning Spain and Holland have sought to carry off from France ; an idea, the consequences of which will soon fill, according to their needs, all the streams of the world ? What, indeed, is this sterile, useless egg, in presence of the millions of fish we can already enumerate, after these few years of almost unaided experiment !

We bring this sketch to a close by a single reflection, but one sad enough : If humanity were in the moon (if we may be pardoned the illustration), and it were possible to establish intercourse with that orb, all our intellectual and material treasures would be lavished upon it. But what should be done for that which surrounds us ? Nothing. Pride would not be satisfied, pride born of egotism, which never produced any thing good. But we may console ourselves by the maxim of immortal wisdom : "Vanity of vanities, all is vanity."

PRACTICAL INSTRUCTIONS

IN

FISH-RAISING.

BY

M. COSTE,

MEMBER OF THE FRENCH INSTITUTE AND PROFESSOR OF THE COLLEGE OF FRANCE.

INTRODUCTION.

HISTORY OF THE DISCOVERY OF ARTIFICIAL FECUNDATION.

When and by whom was artificial fecundation discovered?

When and by whom was the discovery practically and usefully applied?

These are the questions which I here propose to examine.

About the middle of the last century, in 1758, Count de Goldstein, Grand Chancellor of His Palatine Highness for the Duchies of Bergues and Juliers, sent to one of his ancestors of the celebrated Fourcroy, an essay on the artificial fecundation of fishes' eggs, and on the employment of the process for stocking rivers and ponds. This remarkable work, of which Jacobi was the author, being in the German language, M. de Fourcroy found difficulty in translating; Count de Goldstein consequently furnished him with it in Latin. The French version from this was published entire in 1773, in the Treatise on Fish (*Traité Général des Pêches*), by Duhamel du Monceau, by order of the Academy of Sciences.*

I call Jacobi the author, because, many years before Duhamel's work appeared, or the French version of the memoir sent to M. de Fourcroy by Count de Goldstein was known, M. Gleditsch communicated to the Royal Academy at Berlin, which inserted it in the collection of its transactions for 1764, a detailed analysis of the paper by Jacobi, for which it was indebted to Baron Weltheim de Barbke, and which had for its title, "Abridged account of the mode of artificially producing Trout and Salmon, founded upon the practical experience of a skilful naturalist." ("*Exposition abrégée d'une fécondation artificielle des Truites et des Saumons, qui est assurée sur des expériences certaines, faites par un habile naturaliste.*" †) But as the work in question is an abridgment of a memoir in German furnished by Jacobi himself, and as this naturalist proposes in it, *in the same terms,*

* Duhamel du Monceau, *Traité des Pêches*, p. 334. Paris, 1773.

† *Hist. del Acad. roy. des sc. et des belles-lettres*, année 1764. Berlin, 1776, vol. XX. p. 47.

the modes recommended in the copy of the Count de Goldstein, it is clear that to him belongs the honor of the discovery. The original paper in which, *after thirty years of research*, he developes all the consequences of the discovery, was published in 1763 in the *Journal de Hanovre.* The author arrived at these important results in the following manner :

It was known in his time that at spawning season trout and salmon ascend brooks in which clear water runs over a gravelly bottom, choose a resting-place, then working with head and tail manage to move the gravel, and heap it up in a manner to form a sort of dike to break the force of the current, in the interstices of which their progeny may find shelter. In this dike or gravel-bed the female deposits her eggs, pressing her belly against it to facilitate spawning. As the eggs are pressed from her their weight carries them to the bottom, and, as the bottom is pebbly, they fall into the interstices till they fill the bed thus prepared for their reception. Thus placed they are protected from being washed away by the force of the current, and preserved in a state of cleanliness which is necessary to their ulterior development.

It was known too, at the time the memoir was written, of which Count de Goldstein gave a copy to M. de Fourcroy, that the moment the female finishes spawning, the male in like manner pressing with his belly the gravelly bed emits his milt, and that this milt mixing with the current passes like a cloud over the eggs, impregnating them with lifegiving particles, and disappears after having for a moment troubled the transparent water.

Scientific observation had it seems then established, that the contact of the egg and the milt was an *external phenomenon realized between two products of parental organism, expelled from that organism and combining exteriorly to it.*

From this observation of what happens normally in nature to the idea of its artificial imitation, was only a step, and this was plain to the admirable sagacity of the author of the memoir published by the Count de Goldstein. He thus explains it: "If this description of natural propagation by trout and salmon be compared with the artificial processes we have deduced therefrom, we flatter ourselves that in our method will be recognized all the principles indicated as essential to nature." *

According to his description, after having emptied into a vessel a pint of clear water, he seized a female whose eggs were at maturity, and by a slight pressure expressed them into the vessel.

Then he took a male and in like manner expressed into the vessel enough of his milt to give a milky hue to the water in imitation of nature's process, and thus he practised artificial fecundation.

"A pint of very clear water," he says, "is poured into a nice clean vessel, such as a wooden bucket or shallow tub; a female salmon is then taken by the

* Duhamel, *op. cit.*, 2d part, p. 342.

head and held over it, if the eggs have come to maturity they will fall into it ; if not, by pressing the belly lightly with the palm of the hand they can be made to do so. The male fish is then treated in the same manner. When from the male enough milt has been pressed out to whiten the surface of the water, the operation of fecundating the eggs is complete." *

But in order to complete his experiments and turn them to account in industrial application, he had prepared beforehand, to receive the fecundated eggs, long hatching-boxes, in the arrangements of which were combined all the conditions with which he had observed the females surrounded their spawn when deposited at the bottom of streams. He thus describes this hatching apparatus :

"The box may be constructed of any suitable size : for example, eleven feet long, a foot and a half wide, and six inches high.

"At one extremity should be left an opening six inches square, covered by a grating of iron or brass wire, the wires not being more than four lines apart. At the other extremity on the side of the box should be made a similar opening, six inches wide by four inches high, similarly grated ; this one will serve for the escape of the water, the other for its entrance, and the grating will prevent water-rats or any destructive insects from reaching the eggs. The top of the box should be closely shut for the same reason,

* Duhamel, *op. cit.*, 2d part, p. 334

but a grated opening, similar to the rest, six inches square, may be left to give light to the young fish; this however is not absolutely necessary.

"A suitable place should then be chosen for the box, near a rivulet, or, what is still better, near a pond supplied with running water, from which may be drawn by a little canal a stream, say an inch thick, which should be made to pass continually through the gratings and through the box.

"Lastly, the bottom of the box to the thickness of an inch should be covered with sand or gravel, and over this should be spread a bed of stones of the size of nuts or acorns.

"Thus will be made a little artificial brook running over a gravelly bottom." *

It is then in this artificial brook, where, I repeat, are found so cleverly united all the conditions sought by the female in a state of nature, that the author of the memoir published by the Count de Goldstein, deposits the eggs fecundated by the artificial process, the discovery of which belongs to him.

"The eggs thus fecundated are spread," he says, "in one of the boxes so placed, and the water of the little rivulet passes over them, care being taken that it does not run with such rapidity as to displace and carry away with it the eggs, for it is necessary they should remain undisturbed between the pebbles."†

When he had thus scattered the fecundated eggs

* Duhamel, *op. cit.*, 2d part, p. 334.
† Duhamel, *op. cit.*, 2d part, p. 335.

on the gravelly bottom of this artificial rivulet, he watched with care during the six or seven weeks of incubation, all the varied phases of their development, with a view of discovering any hidden obstacle to the success of the experiment. He found that the time necessary for incubation varied with the temperature : that it took a much longer time when the water was cold than when it was moderate. He found, too, that a sediment is deposited on the eggs, which is hurtful or destructive if allowed to remain, and to remedy this difficulty he cleaned them with the feather of a quill.

"Care must be taken to remove from time to time the dirt which is carried by the water and deposited on these eggs ; this can be done by stirring about the water with a quill feather." *

Placed in these favorable conditions, and subject to his assiduous care, the eggs passed safely through every stage of their development. The young fish, thus hatched, seemed quite equal to those hatched in the natural way. After their birth he still preserved them for five weeks, and did not turn them into his brooks till the umbilical bladder appeared absorbed, that is to say, until they commenced to feel hunger.

After an experiment so successfully made, and often repeated, the ingenious author of this discovery had the right to say as he did, that *his method ap-*

* Duhamel, *op. cit.*, 2d part, p. 335.

plied to all species of fish might become a source of great profit. This claim to his discovery, his *work* so well establishes, that it seems singular that any one should attempt to deprive him of it. He places it beyond doubt, by the care he takes to show all the cases where his invention will give results theretofore impossible. Thus he demonstrates the possibility of creating at will, mixed breeds, by mixing the spawn and milt of two different species, which could not be done before his discovery ; he shows, too, the possibility of hatching artificially alongside of ponds containing unproductive species, the eggs of these very species, and of stocking these ponds with young fish from these eggs.

Every part of his research is characterized by such exactitude and practical good sense, that all fundamental questions are resolved ; and this new discovery had hardly appeared in the domain of science when it was transferred to that of industry.

It was in the kingdom of Hanover, near to Nortelem, that the first trials were made. They gave such important results that the fish so obtained became an object of considerable commerce, and England wishing to reward such service, granted a pension to the party who successfully commenced it *

Thus then, not only does the discovery of artificial fecundation belong to the author of the memoir published by the Count de Goldstein, but to him too

* *Soirées helvetiennes*, etc. Amsterdam, 1771, p. 169.

belongs the honor of its application; for the *commercial success*, if I may be allowed the phrase, of the Nortelem enterprise, was obtained under his influence and by putting in practice his ingenious method.

In giving to industry this new method, science too was put in possession from that time of a means of production which she described in all treaties on the history of fishes, and which may be found even in the *Fishing Manuals.* Science has unceasingly reproduced this discovery in her annals, and practised it in her laboratories, in order that, wherever the depopulating of streams was producing want, people might know of a remedy for the growing evil. Science, in now bringing forward the fruits of her past experiences, and regulating their present application to the necessities of the times, only continues to fulfil her mission.

We need not then be astonished, after having read what had previously been done, to find, when in 1837 and 1841, the number of salmon in the waters of Great Britain commenced diminishing, that Mr. Shaw first, and afterwards Mr. Boccius, profiting by the knowledge of the process, formerly so successful in Hanover that their government had rewarded its author, had recourse to the same process for the multiplication of this valuable species. The memoir published by Count de Goldstein, that of Jacobi, and the industrial enterprise at Nortelem, were the antecedents that guaranteed their success.

Consequently, in 1837, Mr. John Shaw * took in the river Nith, in Scotland, a male and female salmon, at the moment the two fish were preparing the bed of stones in which to deposit their spawn, and dug along the shore a ditch into which he turned a current from the stream, and put in it an earthen pot ; then having expressed from the female the eggs into this pot, he expressed from the male the milt into the current flowing over the pot, into which the fecundating particles were carried, and the eggs impregnated.

The eggs thus fecundated were carried to one of the basins he had had prepared for hatching them and preserving the young. There they were deposited on a bed of stones, under a fall of water from a little canal ; and after an incubation, which, owing to the low temperature of the water, lasted a hundred and ten days, the young salmon broke their shells. They were still preserved where they were born, and thrived so well that eighteen months afterwards the males of this brood were capable of reproduction, as was proved by a number of experiments, fecundating with their milt the spawn of full-grown female fish taken from the river.

In 1841, Mr. Boccius, a civil engineer of Hammersmith, carried to still greater lengths practical results of such experiments. He, like those who preceded him, made use of artificial fecundation for

* *Exp. Ob. on the develop. and growth of Salmon fry.* Edinburgh, 1840.

stocking the streams of Mr. Drummond, in the vicinity of Uxbridge, and he estimated the number of trout he produced and brought up there at 120,000. In succeeding years he practised the same process on the estate of the Duke of Devonshire, at Chatsworth, of Mr. Gurnie at Carsalton, and of Mr. Hibberts at Chatfort.*

Just about this time, Mr. Remy, a fisherman of Bresse, an illiterate man, and consequently ignorant that science was already possessed of the discovery of artificial fecundation, the application of which had already produced considerable results; this fisherman, I repeat, seeking a remedy for the decay of his branch of industry, passed many years of his life in one of the most secluded valleys of the mountains of Vosges, in reproducing the experiments already made by so many celebrated physiologists that the long list of their names need not here be recapitulated, and discovering what naturalists had already known for more than a century. Endowed by nature with a remarkable faculty of observation, and with that perseverance that no obstacle discourages, he succeeded in his enterprise. His process of artificial fecundation does not differ from the one described in the memoir published by the Count de Goldstein, discovered by Jacobi, employed by all physiologists, by Boccius—for there could not be two modes of operating, though his hatching boxes are not so rationally

* Boccius, *Fish in Rivers and Streams*, London, 1848.

combined as the artificial rivulets of the Count de Goldstein; rivulets which, notwithstanding their superiority to the boxes of Messrs. Gehin and Remy, have still inconveniences which I have overcome by substituting wicker hurdles or baskets.

But, as Mr. Milne Edwards has already stated in his remarkable report, if scientific men have preceded Mr. Remy in his researches, and if he has not enriched natural history with new facts, he is not the less worthy of our interest, and we owe to him our thanks, for he was the first *in our country* to couple the process of artificial fecundation with the preservation of the brood; a combination which, considering the isolated scene of his labors, had, *for him*, all the merit of a real invention. His first essays, Mr. Gehin having subsequently been associated with him, were made in 1842.*

Notwithstanding the invention having been laid

* The government wishing to make to these two fishermen a proper acknowledgment, granted to Mr. Remy a tobacco factory, and at the instance of the commission (of which I was a member with Messrs. de Suzanne, de Bon, de Franqueville, Mauny, de Mornay, Doyère, and my two brother academicians, Milne Edwards and Valenciennes) an annual pension of 1,500 francs from the budget of the Minister of the Interior; and to Mr. Gehin a tobacco factory at Strasburg, an annual pension of 500 francs, a gift of 1,200 francs, 10 francs a day for travelling expenses while in his department, and 2½ francs per myriameter out of his department. M. Huertier, Councillor of State, Director General of Agriculture and Commerce, who gives his powerful and intelligent co-operation to the organization of fish culture in France, has graciously and warmly responded to these views of the commission.

before the Society of Emulation of Vosges, and the granting by that society of a medal to these two fishermen, it remained buried in their archives ; for it was only in 1848 that the Academy of Sciences was apprised of the claim of the fisherman of Bresse. This claim appeared in consequence of a lecture in which Mr. de Quatrefages, without knowing of the researches of Mr. Remy, called the attention of agriculturists to the fact, that science furnished them a means, discovered a century since, of re-populating their streams, as an organized trade in Germany and experiments in England had proved.

At this period, I had already instituted experiments on the domestication of fish, of which my works on raising eels, and on the nest-building of the Epinoche, are fragments—experiments which I have continued at the College of France, where, as professor of comparative Embryogeny for ten years past, I have been obliged, in elucidating to my class the phenomena of conception, to make them observe the actual processes of nature by the aid of the microscope, and to exhibit to them the experiment of artificial fecundation. The question of the application of this process entered naturally into the habitual course of my studies, and it seemed to me I was giving effectual aid to the organization of a new branch of industry by devoting my laboratory to it. I had in this no idea of setting up a discoverer's claim, or doing aught else than extending to the discovery a benevolent patronage. My life belongs to the new science, the teaching of which has been confided to me, and with which are connected the highest and grandest questions of natural philosophy ; a science of which, thanks to the friendship of M. Guizot, I have had the rare honor of being called to fill the first established professorship.

It is not to be doubted, that in this discovery Germany and England have preceded France, but it is also true, that to our country belongs the honor of establishing its wide-spread popularity and European credit.

Since the day when the results of Mr. Remy's experiments were vouched for by the official report of Mr. Milne Edwards, and that, upon my proposition, it was decided that a model establishment should be founded near Huningen, under the auspices of the government and the direction of Messrs. Berthot and Detzem, Engineers of the Rhone and Rhine Canal ; since, too, people have seen in course of construction in my laboratory, the hatching apparatus which I designed for it ; the public mind has been justly alive to a project touching in so high a degree the interests of society, and constant appeals have been made to the teachers of science to agree upon and make known the best and surest modes of success in this new industrial pursuit.

It is in answer to these calls that I have instituted at the College of France a series of experiments, in which I study, with the greatest care, all the conditions which favor or hinder their success.

Every one will admit, that if the process of artificial fecundation is in the main the same for all the world, its practical details as to bringing up the young fish, their proper food, the mode of transporting them, etc., must, in view of the present diversity of opinions, be established by decisive experiments, which will settle these questions and guarantee success.

To make known this valuable discovery ; perfect its processes ; extend the application of them ; to reduce to certain rules different and uncertain practices ; to introduce all the results which experience constantly gives ; to distribute to all countries, which shall need them for experiments, the fecundated eggs of the establishment at Huningen, in order to excite emulation ; this is the part which, in conjunction with Messrs. Berthot and Detzem, I have taken upon myself in organizing fish-culture.

If I may judge from the evidences of good will which I am receiving from every part of France, and from foreign countries, by the measures taken by the government on the publication of my reports, I may believe that my intervention has not been useless. But artificial fecundation constitutes only one branch of the subject ; with regard to rearing aquatic animals there are others no less important, which I hope soon to succeed in making equally celebrated.

This publication, then, is but the first chapter of the labors I am engaged in, but a response to many demands for information which my occupations have not permitted me to satisfy. I beg of all those who have done me the honor to address me, not to consider my silence as a want of courtesy, but as a necessity of which I now wish to relieve myself on presenting them this work.

PRACTICAL INSTRUCTIONS

IN

FISH-RAISING.

CHAPTER FIRST

PROCESSES OF ARTIFICIAL FECUNDATION.

IF artificial fecundation with the eggs of any species of fish whatever is tried, care must be taken, at spawning time, to bring together in a pond all the fish of that kind intended to be operated upon. Then, having a proper receptacle for the eggs, whether of glass, earthenware, wood, or even of iron tinned, it matters not,—provided it has a flat bottom, as wide at least as its mouth, so that the eggs may spread over it and not lie heaped up,—and having carefully cleaned the vessel, one or two pints of clear water should be emptied into it.

When these preparations are made, a female fish should be taken, and held by the head and thorax with the left hand, while the right hand, its thumb upon the belly, and its fingers on the back and sides, is passed like a ring, lightly, backwards and forwards, and brings the eggs near the opening through which they are passed. (*Pl.* 2, *fig.* 1.)

If the eggs are hard, and already free from the membrane of the ovaries, the slightest pressure suffices to expel them, and under this pressure the abdomen is emptied without injury to the female operated upon ; for the following year she will become as fruitful as if she had spawned naturally, as we have often had occasion to observe at the establishment at Huningen.

If, on the contrary, it appears that a greater degree of pressure is necessary to bring out the eggs, we may be sure they are still inclosed in the tissue of the organ which produces them, and that the operation is premature. In this case it should not be persisted in, but the female should be put back into the pond, and allowed to remain there till her full time is accomplished, care being taken that this will soon occur ; for if a female fish in this condition is kept captive for any length of time in a circumscribed place, her eggs will spoil.

If the females are too large to be held and emptied of their eggs by a single operator, another can aid him in holding them over the receptacle, either by passing his fingers in their gills, or by securing them with a cord, and if the convulsive struggles are very violent, it may be necessary for a third person to hold the tail. The operator, then, with his thumbs upon the thorax and his fingers upon the animal's sides, presses from top to bottom the enormous mass of eggs which distend the coats of the belly. The vertical position in which the fish is held usually suffices to press out the eggs nearest the opening, and the pressure of the hands, repeated several times, will successively bring all the rest.

The easy expulsion of the eggs proves their maturity, for it shows they are detached from the ovaries ; but it does not prove absolutely their capability of being fecundated. For there are some cases, the causes of which we have not ascertained, where the female being in a stream and at liberty, and having gone her full time, and her eggs being ready for delivery, yet she does not or cannot free herself from them, and being thus retained past their time they lose their reproductive faculty.

Experienced persons easily recognize eggs of this sort by two evident characteristics : one is the flowing out with them of a foreign matter, of which there is no trace in their normal state, which gives a muddy hue to the water when the eggs begin to fall into it ; another is, the white color of these eggs when they come in contact with the water. When neither of these appearances is observed, we may be almost sure the operation will be successful ; for the eggs will then be in good condition. But in all cases we must guard against allowing too great a quantity of eggs to fall into one vessel, for if those on the bottom are covered over by too many others, they will not perhaps come in contact with the milt, which should reach every part of them. It will be well, if the females are found to be very productive, to empty the spawn into a number of vessels. The results will then be more satisfactory.

As soon as the process of delivering the female of the spawn is complete, if it appears that the operation of expressing it has brought along with it any part of the mucus which is secreted by her intestines, the water should be immediately changed, so as to free it from every impurity, care being always taken that the eggs are not allowed to become dry. This done, a male fish should be taken, and his milt expressed in the same manner as the female's eggs. If the milt has arrived at a state of maturity it will flow abundantly, white and thick like cream, and as soon as enough has been taken from him to give to the water in the vessel the appearance of whey, it is saturated sufficiently. But in order that the fecundating particles may be spread every where and uniformly, the precaution should be taken of agitating the mixture, and of softly turning over the eggs with the hand, or what is better, with the fine long hairs of a brush, so that no part of their surface shall escape contact with the fecundating element.

After two or three minutes' rest the fecundation is accomplished, and then the eggs, with the water surrounding them, should be emptied into the hatching basins; or if these basins are some distance removed from where the operation has been performed, the water must be renewed before they arrive at their destination, provided the distance be not too great, for then other means must be taken, which I will explain when I come to describe means of transport.

While the mixture is agitated to help the absorption of the semen, if the eggs are of that species which are found to be naturally cemented together by a gelatinous matter, as, for example, are those of the perch, great care must be taken not to pull them apart. This agglutination is a natural condition of their development, of which it would be injurious to deprive them.

There is still another mode of treating the mixture of fecundating particles with the water, which serves as a vehicle, and of aiding their absorption by the eggs to be fecundated : it is to place in the vessel a cullender well riddled, or better still, a fine basket. Into this, while in the water, the eggs are expressed, and then the milt. The cullender should then be moved about, up and down, and from side to side, care being taken to keep it always in the water. This movement has a double result : it thoroughly mixes the fecundating liquor and brings it in contact with every part of the eggs, and the experiment will be successful if, after the agitation of the cullender, it is allowed to remain at the bottom of the vessel quietly for two or three minutes.

A third process is to express into the vessel the milt, and not cause the eggs to fall into the water till it has been thus first charged with fecundating particles. The medium being thus prepared before-band, the eggs reach it in a condition of peculiar aptitude for absorption, which they possess in the highest degree the first moment of their immersion. This mode then seems to offer the greater chance of success. I do not mean to assert that eggs laid in the water some time before the milt is brought in contact with them, lose the power of receiving its influence. For, many times, on the Rhine, I have had occasion to observe that those of the salmon and trout that had been expressed into the water nearly two hours before a male could be caught, still preserved their aptitude for fecundation. But still it is an unfavorable condition, in which, if possible, they should not be placed ; above all, when the eggs of other species are treated, which have not, like the salmon and trout, a protecting and resisting envelope, but which are more sensitive to the influence of the exterior world.

Another mode of treating artificial fecundation, and one more nearly resembling nature's processes, is to spread the eggs on a sieve fitted in a channel or trough of wood or stone, through which runs a current from a water pipe, under the spout of which the end of the trough is placed, and then to pour at this point the spermatized water, and leave to the running current the care of carrying the vivifying particles to the eggs ; but to operate in this way requires an apparatus not always at hand, and perhaps only to be found in an establishment designed for the business. For general use and ready application I recommend therefore, the process described at the commencement of this chapter.

The milt of a single male will suffice to fecundate the eggs of a large number of females, provided he is fed while in the pond or tank, and that care is taken not to take him from the brook and shut him up there until his milt is fully matured. Of this fact the author of the memoir published by the Count de Goldstein was aware, and I have often had occasion to verify it while on board the boat of the fisherman Glasser, at Bale, where the male salmon and trout emptied one day to fecundate the eggs destined for the government establishment at Huningen, are found gorged the next, and so on every day, for the five or six during which their organs secrete semen. It is not necessary, therefore, in experimenting on a large scale, to have numerous males, but only that they should be in the condition I have indicated.

CROSSING BREEDS.

Artificial fecundation gives the means of obtaining, by crossing breeds, mongrels having the qualities of the parents of the two kinds crossed. It will be curious to note all the results from experiments of this nature. At present we know that trout and salmon can be crossed. Trout's eggs fecundated with Salmon's milt, by the operations of Messrs. Berthot and Detzem, on the Rhine, have been sent to me, and have been hatched in my laboratory in Paris. Salmon's eggs fecundated by trout's milt, have been hatched in like manner. It now remains to be determined if the experiment of crossing these two kinds will, in like manner, succeed with all others of the same family; and, also, whether other kinds belonging to a different family, the pike, for example, mixed with the first, can procreate together.

The difficulty which the author of the memoir published by the Count de Goldstein encountered was, that the pike spawns in April, while the salmon and trout spawn in November and December; but now, as we can procure the eggs or the milt of the salmon of the Danube towards the end of April, that obstacle to experiment is removed. We can try the fecundation of eggs of this salmon with the milt of the common ombre, and of eggs of the ombre with milt of salmon, and of eggs of both with milt of the pike, etc.

The Chinese produce these results with golden carp, of which they procure infinite varieties; but their trade consists in confining together the varieties of the same species, and allowing them to produce cross breeds by natural propagation.

CHAPTER SECOND.

HATCHING APPARATUS.

A century ago Jacobi, author of the memoir published by Count de Goldstein, recommended the spreading of the fecundated eggs among the pebbles of the gravelly bed of long wooden hatching boxes, grated at the ends, in imitation of the natural method of spawning practised by the female fish. With this method he had complete success, and its application is still continued in Hanover, where it has so lowered the price of trout as to make that fish a very common article of food. This method is the one which has now been put in practice in France by the two fishermen of Bresse, who, instead of long boxes grated at the ends, have used circular ones riddled with small holes. But modes which seem suitable for experiments on a small scale, or in the beginning, are found inapplicable or inconvenient for a large and well organized trade. Some of these inconveniences are so striking, that I need only name them to show the necessity of recourse to other modes.

In the first place, the dispersion of the eggs among the gravel and shutting them up tightly in boxes, prevents that care which could be given to them if they were always accessible.

Next, the sediment deposited by the water,

whether around or upon the box, or within it upon the eggs, forms soon a thick bed, which oftentimes destroys the eggs. I have seen boxes deposited in the streams at Versailles, of which all the holes were so stopped up with a calcareous deposit, that the water could not penetrate them, and when opened, the contents were found to be decomposed.

Lastly, the difficulty experienced after the young fish are hatched to extricate them from their inaccessible bed without wounding them, is an obstacle to their transport to the ponds where they need to be placed to mature.

These inconveniencies have led us to seek modes the employment of which would enable us, when needful, to handle these products with as much facility as we would inert matter. They are so simple, and of such evident utility, that they must be adopted when explained and understood. They can as well be applied to a regular trade as to the experiment of a laboratory,—to an enterprise on the largest scale as well as to the stocking of a pond or rivulet.

On willow hurdles (*plate* 2, *fig.* 2 *a*) or flat baskets, in our hatching streams, we place the fecundated eggs. Their fine meshes form a sieve through which passes the sediment of the running water underneath, but near the surface of which they are placed. In this superficial position, they can be so easily observed that nothing will escape a careful guardian. If he finds the current so strong as to displace and heap them up, he will moderate it. If any hurtful sediment begins to accumulate on them, he removes it with a fine brush. If after being left some time, a sort of coating of vegetable matter should be formed about them, upon the hurdle or basket, he empties them from this dirty one into a clean one, and by means of this easy transfer, which he effects without injury even to the eggs just hatching or the fish just hatched, he maintains the cleanliness necessary to their development.

But these are not the only reasons for preferring the hurdles just described to the pebbly bottoms recommended by Jacobi and by the two fishermen of Bresse. There is another no less worthy of consideration : it is that after the birth of the salmon or trout, and we speak from having already made the experiment, the hurdles will serve as light rafts on which to float, through a channel of communication, the young fry to the pond in which they are to be kept. To do this, it is only necessary to inclose them in a floating frame which the current will carry to their destination. The crowds of people attracted by curiosity to the College of France have had the opportunity to see more than 10,000 newly hatched salmon or on the point of being hatched, lying at the same time on the hurdles of a simple apparatus of not more than a square metre of surface. This result, obtained in such restricted space, gives some idea of what may be done on a large scale, and for a regularly organized trade.

I give here a figure (*plate* 2, *fig.* 2) representing

this hatching apparatus, in order that landholders who desire to stock their streams, may have similar ones made. It is formed of several small parallel canals, disposed in steps on each side of the principal one at the top, which supplies them all.

After having furnished each one of these canals with a willow hurdle, secured about an inch below the surface of the water, the machine is placed under a cock, so that the water, when turned on, will run into one end of the highest canal. The current will run to the other end, and there find two lateral openings, through which it will flow, right and left, in two little falls, into the next two canals. Through the length of these it will flow in an opposite direction, and find at the other end openings through which it will again fall into other two canals, on a lower plane, and thus from fall to fall, it may pass through any number of compartments, or artificial streams, we may require.

When the machine is in operation, we deposit on each of the hurdles placed in these artificial streams, the eggs to be hatched, the different kinds of which may be separately placed in the different compartments. The current must continually flow over them of the depth of an inch, which is sufficient to hinder the formation of byssus, a kind of vegetation which often destroys them, and from which in this way it is easy to free them as step by step all their changes are noticed.

By these artificial means eggs are developed and

hatched more speedily and surely than in places where the females deposit them, because they are preserved from all variations of temperature and accidents that can retard, alter, or destroy them. Results already obtained, after several years experience in hatching trout and salmon, leave no doubt of the efficacy of this process, or of its capability of adaptation on the grandest scale; for with some slight modification, the machine I have described has served as the model upon which was constructed the vast apparatus of the establishment of Huningen. To operate with it is neither difficult nor expensive; it can be used in a laboratory or on a farm, almost without supervision. All that is needed is a little jet of water flowing continuously.

If the machine I have just described seems too complicated, one can be made like a single one of its compartments. I have several times used a simple wooden box, long and narrow, lined with zinc or lead, or even an earthen fish kettle (*plate* 2, *fig*. 3.) The eggs I have placed on the hurdles with which this wooden canal or fish kettle was furnished, have hatched in my laboratory as well as at the Chateau d'Osman, in the department of Orne, the waters of which seemed to me well adapted for bringing up salmon. A very fine jet will suffice to feed the current, and when the fountain is emptied, it may be filled again, and the operation be thus carried on with little difficulty. It is not even necessary to have always fresh water, for that once used and received in a tub or tank, may be employed again and again, provided it is freed from impurities each time by passing through a filter in the fountain.

If one has neither a greenhouse, orangery, nor coach-house, in which to put up such an apparatus as I have described, and it is desired to hatch eggs in a natural streamlet, it may be done by procuring willow hurdles or large flat baskets resembling the one in the engraving (*plate* 1, *fig.* 4). This basket should be so fastened to the bank that its top should be a little below the surface of the water. The fecundated eggs should then be spread over the bottom, either upon the willow or upon living aquatic herbs covering the bottom.

A willow lid (*plate* 2, *fig.* 4 *a*), or any other suitable cover, should then be fitted to this basket, which will protect the eggs and young fry from their natural enemies and destroyers, and this done, the hatching will take place without any further supervision.

If in the locality where the experiment is to be tried, there are not at hand either willow hurdles or baskets, or the means of making them, recourse may be had to shallow wooden tubs, pierced with a number of holes, covered with wire grating, through which the water may freely flow over their contents. The eggs being placed in them, the tubs should be secured near the surface of the stream.

One of the chief difficulties in the way of natural reproduction is the absence from most streams of shallow shelving banks, with aquatic plants, offering the security which fish seek for depositing their spawn. So too in ponds, the sides of which are evenly dug out by the pickaxe, the water in which is every where of an equal depth, many of the commonest species either are barren or reproduce with great difficulty. The employment of baskets or tubs in a measure supplies such deficiencies, giving the means artificially of placing the eggs near the surface of the water, exposed to the rays of the sun, and, in fact, in a similar position to that which the instinct of the animal would choose as best suited to their development.

The absence of running water sometimes prevents reproduction by certain species, though the animals live and prosper, and even spawn. Trout kept in ponds will spawn, the males will fecundate the eggs, but the eggs fall to the bottom and soon perish. The female labors in vain to keep them clean ; with all her care she can only do so for a very few days. They become covered with dirt, and the germ is destroyed unless laid in gravel and washed constantly by running water.

Hatching apparatus overcomes these difficulties. It is only necessary to procure, by artificial means, the eggs and the milt of even such as are unproductive ones in ponds, and by the aid of such apparatus young fish can be obtained to restock them.

CHAPTER THIRD.

DEVELOPMENT AND HATCHING OF THE EGGS, AND CARES THEY REQUIRE DURING INCUBATION.

A FEW moments suffice for a very appreciable change which takes place in eggs artificially procured and fecundated. The contents seem troubled, become more opaque than at the moment of expulsion from the female, and then insensibly retake their transparent hue ; but in the mean time a little spot, not pre-existing, begins to show itself on a point of the globe in the interior of the egg. This is caused by a coalition of granules, forming what is termed the *germ*, and the coalescence of oleaginous molecules which form about the germ. This modification, which has been erroneously supposed to be the *certain proof of fecundation*, takes place as well with those eggs submitted to the action of the milt, which become impregnated, as with those that do not : at first both kinds appear exactly alike ; except that in the unimpregnated the phenomena is accomplished a little more slowly and irregularly.

But if to the naked eye there is no difference in appearance for the first few moments, or even for the first few days, all doubt will be removed by a recourse to magnifying instruments.

The barren eggs deteriorate rapidly, become more

and more opaque, turn white, or else preserve their transparency, but show no interior change. I have seen numbers of eggs of salmon, trout, ombre, and pike appear thus up to the moment when other eggs preserved with them, of the same spawn, and submitted to the action of the same milt, had come to maturity and hatched.

On the other hand, in eggs vivified with the fecundating molecules, one may see after a time, which varies according to the species and to the temperature of the water, on the interior globe a line, which covers about a quarter of its circumference. This line, which seems whitish when the eggs are on a dark ground, or opaque when they are held up to the light (in the manner in which our farmers examine hens' eggs), is the origin of the fœtus, and represents the spinal column. As this line increases in size, one end of it grows out to a point to form a tail, and the other extends in the form of a spatula. This last corresponds to the embryo's head, and of this there is soon no doubt, for the eyes now appear, two points of a blackish brown, easily distinguished, and forming nearly two thirds of the whole mass of the head. As each day developes its form, the young fish may be seen under the shell or membrane, stretching itself, and drawing itself up, and wagging its tail. When hatching time comes, these movements, the probable object of which is to weaken or tear the shell, become more active. With salmon and trout there is another sign of the approach of hatching besides the quick movements of the young. The outer envelope of the egg becomes a little opaque, and as it were furfuraceous. With other species with which I have made observations, this sign does not appear so plainly. At last a little opening is made in the shell, and that part of the embryo next the opening comes through it. Ordinarily the tail or the head first appears, but sometimes it is the umbilical bladder.

Whatever part may be first disengaged, more than half the body still remains imprisoned, and the efforts of the young fish are unceasing, till after several hours it frees itself from the shell. This membrane, which has protected its development, *but has not served to form any part of its organs*, being now cast off, either is decomposed where it lies, or is carried off by the current.

Certain kinds, like the pike and the ferrat, begin immediately to range about in the waters where they have just been hatched ; others on the contrary, such as the salmon and the trout, weighed down by their enormous umbilical bladder (*pl.* 2, *fig.* 9), can only move with great difficulty, and remain lying on one side, or even on the bladder itself. Some few attempt to move from one place to another, but soon give up the effort.

The time for hatching is not the same with all species. Some, like the pike, hatch at the end of eight, ten, or fifteen days ; others, like the salmon, take from a month and a half to two months.

Besides, development is more or less hastened, according as the temperature of the water in which they are laid is more or less elevated. Pike's eggs placed in a vase, the water of which without being renewed was exposed to the sun's rays, hatched in nine days; while others of the same spawning, placed in the shade in water constantly renewed took eighteen to twenty days to hatch. It required also twenty days to hatch eggs of the ombre, which, more favorably placed, hatched in twelve to fifteen days. Still greater variations of time appear in the incubation of other species of the salmon family. In running water of a warm temperature, the eggs of salmon and trout will hatch in about thirty days, while the same eggs in a cold stream will take seven or eight weeks. The term of incubation may even extend to a hundred and ten days, as was proved by the experiments made in Scotland, by Mr. Shaw, to which I have referred in the introduction.

During their change the eggs should not be left to themselves; they require, on the contrary, a certain watchfulness and frequent visits, in fact, such care as can be easily bestowed by the aid of the hatching apparatus which I use.

Whether the artificial streamlets, which I propose, be used, or in preference to them any other mode, one precaution should always be taken; the eggs should never be heaped upon one another. Their accumulation prevents a proper surveillance of all of them, and besides may retard or even prevent their development. Another and more serious inconvenience often results : if one of the eggs becomes spoiled and covered with byssus, this byssus spreads to the adjoining eggs, and in a few days reaches all that are contiguous and destroys them. The only mode to diminish the extent or arrest the progress of this evil, when the eggs have not been heaped up, is to remove, at once, from the hatching place all that show the least trace of alteration. If in place of sacrificing, an attempt is made to save them by freeing them, with the aid of a brush, from the vegetable parasites covering them, not only will it be a useless trouble, since the tainted eggs are already struck with death, but the evil will be aggravated by spreading over the healthy eggs the particles of destructive byssus, by the very operation of cleansing.

There are cases where the employment of the brush becomes indispensable and efficacious: as when sediment forms in a thick layer on the eggs, whose presence hinders the development of the embryo. It is necessary, then, to remove such matter by passing lightly over them a fine brush of badger hair, such as painters use. (*Pl.* 1, *fig.* 7.)

Lastly, the intervention of man becomes necessary when the larvæ of insects, abundant in certain waters, attack the eggs : from these enemies he must deliver them.

CHAPTER FOUR.

NURTURE OF THE YOUNG FISH.

AFTER being hatched, the young fish observe a rigorous diet, the term of which, varying with different species, ceases with all when the umbilical bladder disappears (*pl.* 2, *fig.* 10). They feel no hunger until after the nutritive elements contained in this bladder have been absorbed, and while it remains they refuse absolutely all other nourishment. My observations, on this subject, repeatedly made with different species, but principally with trout and salmon, agree in every point with those made by Jacobi. Like him, I have noted that trout do not begin to eat till towards the end of the fourth week, and that salmon do not require food till six weeks after birth.

The knowledge of this fact is not without importance for practical purposes, since it fixes exactly the time when the feeding of the young fish should commence. To furnish them with food before the absorption of the umbilical bladder, for example, on the fifth or sixth day, as Mr. Haxo,* after the method of the two fishermen of Bresse, recommends for trout, would be censurable were it not utterly useless, for two reasons: first, young trout for the first month

* *Fecondation artificielle et éclosion des œufs de poisson.* Epinal, 1852, p. 56.

of their existence possess all the nutriment they need in the umbilical bladder ; and next, they could not take and assimilate other food, if furnished them, their intestines not being sufficiently developed to receive and digest it.

Independently of its uselessness, the practice of furnishing food too soon may be really hurtful, especially if the young fish, artificially obtained, are confined in large numbers within restricted limits. Animal matter furnished them, no matter in how small quantity at first, not being consumed, accumulates day by day at the bottom of the vessel, and in the end becomes corrupt, and thus, as I have seen several examples, becomes a cause of mortality.

No matter, therefore, what kind of fish may be obtained by artificial means, there is no necessity to provide them with food until after the disappearance of the umbilical bladder. That most voracious species, the pike, even, is subject to this law, and I have now under observation some of that species hatched twenty days ago, which, having yet traces of the umbilical bladder, do not seek for food.

With regard to salmon and trout it is not enough to know when the young should be provided with food, but also what kind of food should first be given them. This problem, on the solution of which depends the success of fish culture on a great scale, seems to me to have been resolved by the numerous and varied experiments made at the establishment at Huningen, and at my laboratory in the College of France.

As early as 1849, as appears by a memoir on the domestication of fish and the formation of ponds, I had proved by actual experiments, that very young fish could be fed upon the raw flesh of domestic animals, hashed, and that they would thrive and grow rapidly on this diet. This result, which I had obtained with young eels, confined by thousands in a very small space, I tried with salmon and trout. The muscular, raw flesh of full-grown animals, hashed up and pounded in a mortar till it was the consistency of pap, being given to them, I have observed them seize the isolated fragments and devour them with avidity. I had thus the certainty that this species of food suited them, and that I could bring up in a small space, as I had done with eels, a very large number of newly-hatched salmon and trout.

The operation of reducing muscular raw flesh to minute fibrous particles, small enough to be readily swallowed by exceedingly little fish, is one that demands considerable time ; and the further difficulty of separating and properly dispersing in the water the compact mass of flesh resulting from this operation, led me to seek some more expeditious mode of feeding.

To Mr. Chantrant, under whose charge is the hatching apparatus of the College of France, is due the idea of replacing raw by cooked meat. This substitute has had all the success I hoped for, and has enabled me to feed easily, and at small expense, and to bring up in a space 55 centimetres long, 15 wide, and 8 deep, as many as 2,000 young salmon at once. The muscular flesh of boiled beef, which, by being pounded, grated and cut, is reduced to particles proportioned to the size of the little animals it is to nourish, flesh reduced to that state that it will stick together in a mass but can be with greatest ease separated into the smallest particles, I have found, up to this time, the most suitable food for very young fish just beginning to feel hunger.

Experience shows that this food is better for them than calves' liver cooked, or beef's blood boiled. These substances are not sought by young salmon and trout with the same avidity as muscular fleshy fibre, prepared as I have just described. They prefer even the raw flesh of white fish pounded in a mortar, with which, for two years, Messrs. Berthot and Detzem fed and brought up those confined in the reservoirs of the establishment at Huningen. This flesh of fish, well pounded, breaks up, as does that of boiled beef, into vermiform particles, for which young salmon show great liking.

As for the rest, whichever among these aliments may be the one adopted, if the little fish intended to be reared to the size at which they may be transported, are kept in narrow, artificial streams, or in large vessels, wherein the water though changing has not a rapid current, care must be taken, in order to avoid accidents easy to foresee, to cleanse from time to time, by means of a glass pipe, the bottom of the stream or vessel of the deposit formed of particles of animal matter which the fish have not eaten.

I have thought that this inconvenience, which without great care is followed by serious results, might be easily avoided, if, in place of dead food, the young salmon and trout could be furnished with living prey. Although the results I have thus far arrived at in experimenting with this view are not yet confirmed by long practice, they nevertheless appear conclusive enough to be reported.

The spawn of the frog, so extolled by the fishermen of Bresse, was the subject of my first experiment. The eggs of this animal were placed in ponds in which were hatching boxes containing young salmon and trout, and were there developed and hatched; but neither the tadpoles nor the albumen that enveloped them were sought by the young salmon and trout. I do not mean to say that such prey was not to their taste, but only that its size was too great for such little fish. If tadpoles are suitable food for fish a year or two old, I am convinced they cannot be fit for little ones which have just lost the umbilical bladder. I found it necessary, therefore, to abandon the attempt of feeding fish of that age with such food.

I had recourse to another expedient and with happier results. Pike's eggs, artificially fecundated, being put to hatch in reservoirs in which were very young salmon, the young that came from them were small enough to be easily swallowed by the salmon, and many persons have beheld with me the young salmon darting at them the moment they appeared from the egg. Often they would even eat the membrane or shell of the egg from which the young had been hatched.

I do not hesitate, then, to recommend this kind of food as the one most suitable, most resembling that obtained in a state of nature, and best adapted to the appetite of young salmon. And, besides, it is simple and easy, for it only requires the artificial fecundation of eggs, not alone of pike but of other kinds of white fish of little value, and the hatching of them either in separate boxes, or even in those containing the young they are to find.

Another living prey which young salmon and trout appear to relish greatly, consists of almost microscopic crustacea, of the species cythère, cyprès, and cyclops, etc., which can be found in great abundance, particularly in spring, in all stagnant waters. They can then be taken in such quantities as to serve as the chief food for very young salmon. These microscopic crustacea, always in movement in the water, are for young salmon and trout an attractive food, they seek with avidity and thrive upon.

Lastly, very small, newly-hatched earth-worms, are also a prey for which young salmon and trout have a predilection ; but it is not always easy to procure them in abundance, and they must therefore be considered as an occasional luxury.

Whatever be the regime to which the young fish are submitted, whether living prey be obtained for them, or in default of it they are fed upon cooked meat, in all cases it is possible to bring them up, to the number of many thousands, within very narrow limits, and to cause them to grow rapidly until they attain sufficient size to be let loose in larger streams.

But there the same care and the same provision of food should be extended to them. A perfect system of economy even requires their being furnished with food at all stages of their existence. Not only will this prevent those of a carnivorous species from preying on one another, but they will acquire soon a more beautiful shape and better qualities than they could do if abandoned to the resources which streams naturally offer. Means of feeding become more easy as the fish grow older. Thus, to salmon a year old, can be furnished with little trouble, in great abundance, tadpoles, the fry of white fish, and principally of minnows, aquatic mollusca, little fresh water shrimps. Those of a more advanced age will thrive upon the leavings of the kitchen, and upon the flesh of all kinds of domestic animals.

CHAPTER FIFTH.

MEANS OF TRANSPORT OF EGGS OF FISH JUST HATCHED, AND OF FISH TO STOCK PONDS OR STREAMS.

TRANSPORT OF EGGS.

In China may be seen, towards the month of May, a great number of vessels collected in the great river *Yang the Krang*, to buy there the seed of fish, a custom which has continued for ages. The country people bar the river in many places for the space of eight or ten leagues with nets and hurdles, leaving only space for the passage of a single vessel. The seed of the fish is arrested by these hurdles, when these people perceive it, though a stranger's eye would not discover it. They dip up the water containing the seed of the fish and empty it into large vessels, which they sell to traders, who take it to the provinces, and resell it in smaller quantities to proprietors who own rivers, brooks or ponds, which they wish to stock with fish.

The Romans did as the Chinese, and had recourse to the same means to stock their streams, applying them on a vast scale, sowing eggs as they would grain, and carrying this trade even to the extent of hatching, in fresh waters, the spawn of sea fish, which they thought thus to acclimate. Thus the lakes Velinus, Sabatinus, Vulsinensis and Cimnius, in Etruria, were stocked with barble, goldfish, muges, and all species that could be adapted to this caprice.* The rural descendants of Romulus and Numa practised this mode of breeding as a measure of public utility, which gave them, in their rustic life, an abundance which they carefully guarded. But towards the commencement of the seventh century, when luxury and vanity took the place of the simple manners of this ancient race, fresh water fisheries for the people began to be despised, and in their stead were sea fisheries for the rich.

The transportation of eggs to great distances is, then, a fact, of which experience amply proves the possibility. The only question then is, how to transport them without waste, and in the most economical manner.

The two fishermen of Bresse, Gehin and Remy, put the eggs in a tin box, pierced with holes, like those they use for hatching. They cover the bottom with a bed of wet sand, half an inch or more deep, upon which they spread a layer of pebbles of the size of playing marbles. In the interstices of the pebbles they place a certain quantity of fecundated eggs, and cover the first with a second layer of pebbles ; the interstices of the second layer are then filled with eggs, and then of another and another layer, till the box is filled.

* Columelle, *De re rustica*, b. vii. c. 16.

This process presents defects which prevent me from recommending its application. Besides the exposure of the eggs to breakage between the stones, jumbled together by any shock received by the vehicle transporting the box, the holes with which the sides of the box are pierced, favor the evaporation of the water which keeps the sand wet, and exposes them to become dry.

Very fine wet sand employed alone is far preferable. It should be thus used : Take a circular or oblong box, made of very thin white wood, such, for example, as are used for packing dried fruits (pl. 2, fig. 5). Then, on the bottom of this box, spread a bed of wet sand ; on this sand spread as many eggs as can lie together without touching one another. In fact, leave spaces between them that the second layer of sand will fill, to prevent the possibility of their pressing against one another. Then spread a second bed of wet sand, and on this, in like manner, a second bed of eggs, and so on, till the box is filled entirely, so that the lid will press upon the sand and prevent the movement of the contents of the box.

A suitable box to carry eggs in this manner, should not be more than four inches deep, by eight or ten inches long, for if the dimensions exceed this, the weight of the sand will be too great for the eggs to bear. Partitioned off into compartments, a larger box would answer, but it is more simple and commodious, if the quantity of eggs to be sent cannot be packed in one such small box, to use several such, and to tie the whole together or enclose them in a large basket.

Placed in layers alternately with this wet sand, eggs of most kinds of the salmon family, that is to say of the sorts with shells of a certain toughness, can be perfectly preserved for many days, and even during a month, if the box is kept at a somewhat low temperature. This is proved by the following facts. Salmon and trout eggs, fecundated artificially, were put by Messrs. Berthot and Detzem, at the end of December, 1851, in a fir box filled with wet sand. The box was then *for nearly two months* kept in a cold room, but in which the temperature was never so low as freezing point. After the lapse of that time, the eggs were sent to me from Mulhouse. Before taking them out I dipped the box in water, *so that the sand it contained might be gradually moistened throughout*; for had I neglected this precaution they would have perished, as did others not thus treated. The box being then opened, I found them a little withered or wrinkled; but, placed in the hatching apparatus they soon retook their spherical form, and a large number of them gave birth to young fish. Doubtless there were many eggs that did not hatch; but, under like circumstances, when eggs are sent to a distant place to propagate a foreign species, the number that can be hatched, however limited, will still, as in this instance, suffice for the purpose.

This long sequestration in a box will not answer

with like success in application to species that require only a very short period for incubation; for notwithstanding that in such circumstances, development of the embryo may be repressed, it still is going on to a certain extent, and the eggs may be hatched in the sand before arrival at their destination. With the eggs of salmon and trout this difficulty need not exist, for the time of incubation extends to forty-five or fifty days at least, and is prolonged even to a hundred or a hundred and ten days if the temperature of the water is very low; but with such as hatch eight or ten days after being spawned the same results cannot be hoped for, nor can such kinds of eggs be conveyed to great distances.

Ordinarily, we use advantageously aquatic plants instead of sand: we choose among the plants found in the waters where the fish are taken whose eggs are to be transported, such as are softest and least liable to become matted together. We place in the boxes alternately as with the sand, layers of leaves and of eggs. When the box is filled the lid is closed, and the humidity retained by the leaves suffices to preserve the eggs. For several years past, Messrs. Berthot and Detzem have sent me from the banks of the Rhine, and from the establishment at Huningen, a great number of these boxes, and I have had by these means often as many as 10,000 eggs at a time on the hurdles of my hatching apparatus at the College of France, almost all of which have been hatched. Recourse may be had to either of the processes I have described for packing eggs for transportation, since both modes have resulted satisfactorily.

To be certain of transporting eggs to arrive in good condition for hatching, it is by no means immaterial to select the proper period of their development for packing them. Of this I feel assured from numerous experiments, *at least with regard to the eggs of trout and salmon.* Those which are packed in boxes to be sent a long distance, immediately after they have been fecundated, are much more sensible to destructive influences than those which are packed at the period when the embryo is so developed that through the shell or membrane its eyes may be perceived like two black specks. In fact, this is the best time to transport them; first, because they support the journey more readily, and next, because it is only then that we can know certainly that they really are fecundated. When the establishment at Huningen comes to be able to furnish eggs to every point in France, where secondary hatching apparatus may need them, we shall conform always to this rule for transportation, with the certainty it will be successful. Already we have practised it in several departments.

TRANSPORT OF NEWLY-HATCHED FISH.

The younger the fish are the easier it is to transport them a great distance. Even those destined to live in running water, which require it for hatching, as all varieties of the salmon family, can be preserved a long time in vessels, the water of which is not even frequently renewed ; but for this result they must be placed in them immediately after birth, and in the vessels must likewise be placed living aquatic plants. I have made on this point numerous experiments, which leave no doubt of the efficiency of this method. I have frequently put two hundred young salmon, or two hundred young trout, in a glass jar containing no more than three quarts of water, which, being renewed every three or four hours, I have thus been able to send them great distances, to places where it was believed these animals could not be acclimated, and where they are now thriving.

If, instead of taking the trouble to renew the water from time to time, a continuous stream could be introduced into it, there is no distance, no matter how great, they might not be sent at this first stage of their existence. I have kept in the College of France as many as six thousand at a time, in wooden boxes or earthen vessels not larger than eighty centimetres long, fifteen wide, and ten deep, with a stream of water no larger than a straw. The current made by this little stream was found sufficient to preserve them for more than a month in as sound a state as if they had been swimming in large streams ; and I have them still in such vessels and continue therein to rear them. Now, if the experiments which I have just cited, prove that we can preserve thus long in such restricted space, such a prodigious quantity of newly-hatched fish, without the necessity of furnishing them any thing else than an imperceptibly small stream of water, it is evident that through our rivers and canals, boats suitably arranged could carry them in masses to every part of France, so that there is not a single point at which they may not be distributed.

TRANSPORTATION OF OLDER FISH.

When they have attained a sufficient size to be suitable to stock streams, the young fish are much more difficult of transportation to a distance. Nevertheless, by recourse to boats converted into tanks, they can be thus carried great distances and distributed at all points on the voyage. Messrs. Berthot and Detzem last year dispatched one of these convoys from the government fish-establishment at Huningen, which arrived at Dijon in twelve days, having gone over by land and water a distance of 120 kilometres. Fifteen hundred salmon thus carried arrived in good condition and were deposited living in the basin of the Garden of Plants in Dijon.

If the duration of the voyage requires it, food should be supplied to them ; but care should be taken, in case this food consists of dead animal matter, that the vessel should be kept clean, because the remains of such decomposed matter left in it might prove destructive to them. It would be fitter and more prudent to supply them only with young living prey of such a kind as formerly described.

PERIODS

OF SPAWNING OF CERTAIN KINDS OF FISH WHICH REPRODUCE IN FRESH WATER.

NAME OF THE SPECIES.	TIME OF SPAWNING.
SALMON (*Salmo Salar*),	From November to February.
SALMON HUCH (*S. Hucho*),	April and May
TROUT (*S. Fario*),	From October to February.
COMMON OMBRE (*S. Thymallus*),	April and May.
OMBRE CHEVALIER (*S. Umbla*),	February, March and April.
LAVARET (*S. Wartmanni*),	August, September and October.
FERA (*Coregonus Fera*),	January and February.
SHAD (*Clupea Alosa*),	March, April and May.
PIKE (*Esox Luceus*),	February, March and April.
CARP (*Cyprinus Carpio*),	From May to September.
BREAM (*C. Brema*),	End of April and May.
GIBELE (*C. Gibelio*),	May, June and July.
TENCH (*C. Tinca*),	June and July.
PERCH (*Perca Fluviatilis*),	March, April and May.

Note—The periods indicated in this table, varying, according to places and climates, must not be considered as fixed, but as terms, considering which it is possible to guess pretty nearly the times at which the eggs of the different species will be likely to hatch by artificial means.

In the year 1850, the attention of the French Government being called to the discovery by the two fishermen, Gehin and Remy, of a mode of artificial fish culture, the Minister of Agriculture and Commerce appointed a member of the French Academy, a distinguished *savant*, M. Milne-Edwards, to examine into the whole subject, and make a report.

The following is a translation of the document submitted by him :

Report on artificial Fish-culture, and on stocking barren or impoverished rivers with fish, artificially hatched ; made to the Minister of Commerce by M. Milne-Edwards, member of the Institute.

SIR :

Owing to the interest which you feel in all discoveries calculated to increase the alimentary resources of the country, you desired to form a correct opinion of the attempts which, for some time, have been made, whether in France or in England, to ensure the multiplication of fish, in ponds and rivers, and to augment the value of products of fisheries.

You have done me the honor to submit this question to my examination, and have charged me most particularly to render a complete account of the results obtained by two fishermen, who followed their trade near the sources of the Moselle, and who, by a process of artificial fecundation, have established in the department of the Vosges, a veritable *fish factory*. With pleasure I conformed to your wishes, and I will be well pleased, Mr. Minister, if the investigations I have made, can aid you in endowing our rural industry with a new source of wealth, the importance of which will not be undervalued by physiologists or agriculturists. Fish is an article of food, rich in nutritive qualities, and to augment its abundance, either on our coasts or in our streams, will be a real benefit for all classes of population. River fishing is generally little productive in France ; but it is only necessary to cast one's eyes upon the doings of our neighbors of other countries, to comprehend what might be its value, if means be found to stock with good fish our rivers and ponds, as amply as nature has stocked those of Scotland and Ireland, and as agriculturists stock their fields with herbivorous animals, equally destined to serve our subsistence.

River-fishing has long been the object of enactments favoring the reproduction of fish, and protecting the development of the fry. The royal ordinance of 1669, forms the basis of our legislation on the subject, and contains many clauses of incontestable utility.

Proprietors of ponds bestow ordinarily some care upon stocking them, but all that relates to reproduction of fish in our rivers is left to mere chance, and while bitterly lamenting the constant and rapid decrease of their products, we have not, till now, given sufficient consideration to the remedies for the evil.

Public attention was at last awakened to this question, by a lecture delivered two years since, at the Academy of Sciences, by one of our most distinguished zoologists, M. de Quatrefages, formerly one of the Faculty of Science of Toulouse. This learned and elegant writer, gave our agriculturists useful counsel on the art of bringing-up fish, and strongly urged upon them the putting in practice a process of multiplying their numbers, long well known to physiologists, and often experimentally employed in their cabinets, namely, that of artificially fecundating the eggs. We know by the labors of Spallanzani, and by the experimental researches with which you, yourself, Mr. Minister, and your ancient colleague, Prêvost (of Geneva), twenty five years since enriched science, that all fecundation is the result of the action exercised upon the egg at its state of maturity by the living spermatozoa with which the semen or milt is charged; that this action takes place through the direct contact of these two reproductive elements, and that the physiological puissance of these same agents may be preserved during a longer or shorter period after they have been taken from the living bodies which have given them existence.

With a great number of inferior animals the parent's part in the work of reproduction, consists only in the formation and emission of these two generic elements; the egg is not impregnated till, after being spawned, it meets the spermatozoa, the contact with which, necessary to endow it with life, only takes place by the concurrence of exterior causes, independent of the action of the parents, for example, by the course of the current in which the milt is deposited. The experimentalist can, therefore, determine at will this physiological phenomenon, by mechanically mixing the eggs and milt of these animals, and the same results will be obtained by this process as by the natural one.

The observations of zoologists show, too, that in the general harmony of nature, the fecundity of animals is regulated, not only with regard to the causes of destruction to which the young are exposed before they become capable of reproducing their species, but also in view of the chances of nonfecundation to which the eggs are submitted, as the contact of the eggs with the seminal fluid takes place after they have been spawned and depends more or less upon chance. Fish belong, for the most part, to the category of animals among which there is no act of copulation for reproduction, that being effected simply by the ejection by the male of the milt or semen upon the eggs which have been spawned by the female.

To procure the development of the embryo, therefore, in the otherwise sterile eggs, the naturalist, in the experiments of his laboratory, has only to imitate that which happens normally in nature; that is to say, to bring them in contact with water charged with milt; impregnation, then, is soon effected, and to procure this milt, as well as the eggs to be impregnated, all that is required is a light pressure of the abdomen of the males and females, whose products are matured and whose lives will not be endangered by the operation: or these products may even be procured by opening the bodies of newly dead subjects, for the eggs and the milt preserve their vitality for some time after the death of the bodies containing them, and thus from two corpses may be brought forth a numerous and strong generation.*

* This report of Mons. Edwards I have translated from a work entitled "*Fecondation artificielle et éclosion des œufs de poissons, suivi de réflexions sur l'ichtyogenie, par le docteur Haxo, d'Epinal, Secrétaire perpétuel de la Société d'Emulation des Vosges, Membre de la Société des Sciences, Lettres, et Arts de Nancy, etc.*" The object of Dr. Haxo's work appears to be to prove the claim of the two fishermen of Vosges, Gehin and Remy, to the title of discoverers of the method of artificially impregnating and hatching fishes' eggs, by showing that they were the first to bring it into use and practically prove it could be done, and that all others had only theoretically treated the subject till these two poor fishermen took it in hand and showed to the world its value. Dr. Haxo claims it for them as their original discovery, on the ground that they were so unlettered as to have been utterly ignorant of any researches or experiments of naturalists. He insists that they have been badly treated, their discovery stolen from them by naturalists who have no right to it, yet claim it as their own, or as belonging to discoverers of a past century. Dr. Haxo brings several documents to fortify his position, and comments with great warmth upon the injustice towards these fishermen displayed by M. Coste, in the work of which a translation forms part of this volume, as well as by M. Edwards in this report. As the details of the method pursued by Gehin and Remy, given in Dr. Haxo's work, are not fuller than those here given from the pamphlet of Godenier, and as his work seems for the most part a defence of the claims of Gehin and Remy as discoverers, I have not thought that a translation of it would add to the value of this volume, as a manual of fish culture. Dr. Haxo accompanies the publication of M. Edwards's report with many notes to show its injustice to the two fishermen; as a specimen of them I quote this one referring to the passage above.—*Translator.*

"After reading this passage who is there who would not be led to believe that the processes of artificial fecundation were not perfectly known, at least by savants? But notwithstanding this, M. de Quatrefages says not a word of them in the memoir he presented

This fact was fully established by Count de Goldstein, about the middle of the last century, long before Spallanzani published his beautiful researches upon generation. In 1758 this judicious observer addressed to an ancestor of the celebrated Fourcroy, a most interesting memoir upon artificial fecundation of trout's eggs, and upon the application to stocking rivers of which the discovery was susceptible.

An extract from Goldstein's work was inserted in a work called *Soirées Helvétiennes*, and some years later, in 1770, Duhamel du Monceau gave a translation of it in the third volume of his *Traité général des Pêches*, published under the sanction of the Academy of Sciences.

to the Institute in 1848; while on the other hand, when the letter which I addressed to that learned body on the 2d March, 1849, was read by M. Flourens it was received, according to the testimony of the Abbé Moigno, who was present at the meeting, with the most unequivocal demonstrations of surprise and satisfaction on the part of all the members of the Academy of Sciences. M. Milne-Edwards was then immediately appointed as one of the commission to examine my report in conjunction with Messrs. Duméril and Valenciennes. How does it happen that he did not then inform his colleagues that the matter had been long before known? How was it that he did not then and there announce that not only the processes of artificial fecundation had been very many years before described by Goldstein, by Duhamel du Monceau, and by Jacobi, but that they had been successfully practised in Scotland? Why did he wait, before making any such statements, until after he was officially charged by the Minister of Agriculture and Commerce, *to go to the place and examine the results of the labors of the two Vosgian fishermen?* We leave all such reflections as these to the sense of the reader."

About the same period, a German naturalist, Jacobi, published at Hamburg an equally interesting letter upon the art of bringing up salmon and trout, and on the production of these fish by means of artificial fecundation. At a later date analogous experiments were made in Scotland by Dr. Knox, Mr. Shaw, and Mr. Andrew Young. In 1835, Signor Rusconi, so well known among naturalists by his work on the embryology of salamanders, published in the seventy-ninth volume of the *Bibliotheca Italiana*, new observations on the development of fish, and gives equally instructive details in artificial fecundation of the eggs of the tench and the ablette. At my suggestion, the translation of this memoir was inserted in the *Annales des Sciences Naturelles, pour* 1836.

I would add, too, that it was by recourse to this method of multiplication that Messrs. Agassiz and Voght procured all the embryos necessary for their studies on the development of the palée, a species of salmon of the Swiss lakes, the anatomical history of which these two naturalists published in 1842. The philosophical fact, then, upon which M. de Quatrefages relied to stimulate agriculturists to the *manufacturing* of fish, in the same way they produce grain or meats, offered nothing new to zoologists, and to their remembrance M. de Quatrefages was the first to recall the claim of Goldstein as the discoverer of artificial fecundation. But under our system of education, truths well known by naturalists are unknown by most other men even the best informed, and it was not unnecessary to call public attention forcibly to this application of science to rural industry, which not only had not profited by the results of the discovery, but I think I can safely affirm that there were then not ten agricultural authors or teachers in all France who had the least idea of the service which physiologists had so long before rendered them.

Under such circumstances we should not be astonished to find in one of the most secluded valleys of the chain of Vosges, two illiterate fishermen, but endowed by nature with a rare spirit of observation and a rarer perseverance, being ignorant of prior discoveries, and wishing to find some remedy for the decrease and threatened extinction of their trade, employing several years of their time in laboriously making over again the same experiments already made by the physiologist I have cited, and in re-discovering what naturalists had been acquainted with for a century.

But if these poor peasants of Bresse were preceded in their researches by scientific men, and if they have not enriched natural history with fresh discoveries, their labors are no less worthy of interest, and they have a claim upon our consideration, for they seem to have been the first among us to make practical application of the discovery of artificial fecundation to the rearing of the fish, and have thus the merit of creating in France a new branch of industry.*

The first essays of Messrs. Gehin and Remy were made in 1842. Having by a long course of observation become acquainted with the mode of reproduction practised by trout, and being assured of the possibility of artificially fecundating its eggs, they applied themselves to the production of quantities of these fish to stock the streams of the canton. Success crowned their efforts, and notwithstanding their feeble resources, and the difficulties of all sorts they had to encounter, they still obtained considerable results.

They were enabled to stock, with young trout artificially hatched, two ponds near their village of Bresse, one of which furnished last year 1200 trout of two years old.

Gehin and Remy estimate at about 50,000 the number of young fish they have put in the Moselotte, a little river of Bresse, which empties into the Moselle near Remiremont; they have put in practice their mode of stocking in several other localities of the same canton, as appears by documents furnished by the authorities of Saulxures, of Cornimont, and of Gérardmer. Besides these, M. Kienzi, Mayor of

* M. Edwards, in this paragraph of his report, seems to concede to the two fishermen all that can be claimed for them as discoverers of a mode of rendering a discovery useful and valuable; notwithstanding Dr. Haxo's pamphlet to prove the injustice of M. Edwards and others towards them.—*Translator.*

Waldenstein, in the department of Haut-Rhin, deputed them to re-stock the water-courses of his commune, and this intelligent official gives assurance that they perfectly succeeded.

I would add also, that, wishing to render the discovery of the widest public utility, our fishermen never made any secret of their processes, but, on the contrary, readily initiated any one who desired to undertake similar work. All who have ever had occasion to witness the labors of Géhin and Remy bestow on them the highest praise.

I visited their establishment and witnessed some of their experiments. The Society of Emulation took up and fully investigated the subject, and bestowed on each of these worthy men an honorary medal. The work they proposed it seems to me they fully succeeded in, and to render their country great service they only need the means to extend their operations. I do not judge solely by the results obtained by Géhin and Remy, but also by similar ones on a large scale, which I found to have been obtained for several years past in Great Britain, and which had excited there considerable interest.

In fact, M. Boccius, a civil engineer of Hammersmith, has practised artificial fecundation in stocking several rivers of Great Britain, and seems to have had complete success.

In 1841 he worked in the streams belonging to Mr. Drummond, in the neighborhood of Uxbridge, and he estimates at 120,000 the number of trout he there brought up. The following years he put in practice the same processes on the magnificent domain of the Duke of Devonshire, at Chatsworth; then for Mr. Gurnie, at Carsalton; and Mr. Hilbert, at Chatford; finally, the Angler's Club put under his charge the important fishing-ground of Ansval-Magna, in the county of Hertford, and M. Boccius assured me that he had already artificially hatched there at least 2,000,000 trout. He has published a book upon his method of stocking streams, and it seems that a society is about to be formed, under the patronage of Sir H. Labouchere, with a view of attempting to stock the Thames with salmon.

The process employed by Géhin and Remy is very simple and easily practised; it hardly differs from that adopted by Boccius, and equally resembles the method described by Jacobi, nearly a century ago.

Trout-breeding takes place in December, and in order to have eggs for artificial hatching, it suffices to press lightly, before and behind, the abdomen of a female fish ready to hatch; and her eggs in falling should be caught in a vessel with water, and afterwards sprinkled with milt obtained in the same manner and diluted.

If the eggs have not arrived at their term when operations are commenced, they will only be run out with a strong pressure, and in such case the fish should be left in a preserve during some days, before this forced birth is adopted, for neither the eggs nor the milt can be usefully employed in a state of immaturity, and the life of the parent fishes would be endangered by rough handling.

On coming in contact with the spermatised water, the eggs change color: before fecundation they are transparent and yellow: so fecundated they become whitish or rather opaline. A trout aged some two years* and weighing about 125 grammes, can furnish about 600 eggs; a trout of three years, 700 to 800; and it is also to be noted that the milt of one male is enough to fecundate the eggs of a half-a-dozen females or even more. Messrs. Géhin and Remy placed the eggs so fecundated in a tin box pierced with holes on a gravel-bed: these boxes are about fifteen centimetres in diameter, and eight deep, and can contain each a thousand eggs. They are then to be placed in some streamlet of which the waters are pure and lively, but not deep: in this they are partially buried, and so disposed that the water in the boxes is rapidly renewed, for the agitation of it is necessary to assure the respiration of the embryos, and also to hinder the development of confervas, which will not be slow to catch and destroy the eggs if the water be stagnant. The development of these embryos lasts four months, and it is generally towards the end of March or in April that the hatching takes place; during six weeks more, the new-born trout carry under the abdomen

* Experience shows that the trout does not become nubile or fit for propagation before the age of three years.

the umbilical vesicle which holds the remains of the nutritive matter, analogous to the yolk of a bird's egg, and at first by means of this substance the minnows are nourished ; but when absorption takes place the young fish have need of other nutriment, and should then be driven out of the box in which they are cradled, and permitted to swim freely in the streamlet which they are to stock.

In fine, to procure for these little fish suitable and abundant nourishment, it is only necessary to leave or put in the water some frogs, whose spawn they will greedily eat, while the tadpoles afford excellent food for the older trout. When the young trout so brought up, are destined to stock a river, they should be placed in streams tributary to it, and water selected which rushes over pebbles or rocks.

In proportion as these fish grow, they descend spontaneously to the deep water, whither they arrive only when they are sufficiently agile to protect themselves against the enemies which they may encounter ; while if they are at once placed in the midst of other voracious fish, they will have but a small chance of escaping death. When they are so raised in streamlets or ponds, precaution must be taken to separate the product of each year from the former one, as the big trout will otherwise eat up the little ones ; and to avoid this the young fish in the same circle should be of one age.

To establish after a regular fashion this branch of production, there should be at least three streamlets or brooks, for the fish to be changed during three years, new ones being placed in them as fast as exhausted.

Unhappily Messrs. Géhin and Remy have not at their disposal the necessary funds to complete this work. They have obtained the grant of a fish-pond for this purpose, and bought another for 800 francs; but now their pecuniary means are gone, and if, sir, under your kind protection, they do not get some help from Government, I fear it will be impossible for them to pursue the trials so satisfactorily commenced.

The labors of Messrs. Géhin and Remy appear to me the more worthy of encouragement, as success can afford but little profit to such devoted and active men, but will contribute to increase the alimentary resources commanded by the people on the banks of streams. Only in considering fisheries as works of public utility and causing them to be executed by the state, can we hope to give real importance to our river fisheries; but in applying a small sum to this end, we will arrive, I have no doubt, at important results for the country.

If the fish-breeding practised by Messrs. Géhin and Remy were only applicable to trout, and to other fish of limited supply, I would not attach as much interest to it as I do; but it may be applied to salmon, and I am convinced that it would be easy thus to restore to the rivers of Brittany icthyological riches which are now disappearing, and even to acclimate salmon in rivers, which, up to this time, have not been frequented by that fish.

Nothing is easier than to transport eggs just laid,* or living salmon, of which the abdomen is filled either with eggs or milt; and even when these die on the road, the hatching of their eggs can be attained. In placing the eggs so acquired in streamlets properly chosen, the young salmon will grow as though spawned there by their parents; they will emigrate as usual to the ocean, and in its depths they in turn will spawn, and will not fail to return in great numbers to the stream whence they proceeded, and in following its course seek a proper place for the growth of their progeny.

We know, in fact, by experiments already old, made in Brittany by Delandes, and by observations of the same kind, repeated in our day, in Scotland by the Duke of Athol, Sir W. Jardine, Mr. Baigrie, Mr. Hayshan and Mr. Young, the Director of the fisheries of the Duke of Sutherland, that guided by a singular instinct comparable to migratory swallows, the salmon after having emigrated far into the sea, returns ordinarily to the water where it was spawned, and the individuals of the same species are so perpetuated in certain rivers without mixing with those of strange waters. It seems to me consequently indubitable, that in the space of a few years, it

* With due respect to M. Milne Edwards, the transport of such eggs is very difficult, and if this difficulty has finally been obviated by Géhin it is only after much groping and research.

would not only be possible greatly to multiply salmon in all the waters natural to them, but to introduce and acclimate this large and valuable fish, in many of our streams hitherto without them. For the salmon and the trout also, as well as for many other kinds, the method of Géhin and Remy appears to be the surest mode of stocking rivers; but we cannot have recourse to the artificial fecundation of eggs to stock fresh waters of certain kinds, of which the introduction, however, would be of great utility in certain localities. Thus, eels are never caught at maturity with milt or eggs, and these fish seem to be only produced in the depths of the sea, whence just spawned they go in legions innumerable every year, to occupy rivers, where they are known by our fishermen under the name of *montée.*

To supply brooks and streams needing them, such spawn must be transported, and the operation renewed periodically; and M. Coster has shown that this transportation can be easily effected, even to considerable distances.

For this purpose it is sufficient to place the young eels in grass kept wet. The experiments which M. Coster is now pursuing at Paris in the laboratory of the College of France, proves that young eels can be fed at small expense, so that they will grow rapidly, and it seems to me that in many marshy places, raising eels would be profitable.

If I had to treat here of marine fishing I would ask of you, sir, permission to call your attention to several matters touching the treatment of our oyster-beds, and the means of favoring the multiplication of these molecules. A manufacturer of Charente, M. Carbonnel, has conversed with the Academy of Sciences several times lately, and thinks it would be easy to establish on our coast at different points such artificial oyster-beds. M. de Quatrefages has also requested the naturalists on our coasts to try the artificial fecundation of oysters, and I am persuaded that in studying experimentally all that relates to the generation of these molecules, we shall arrive at results extremely interesting for industry as well as science. But in the actual state of our knowledge relative to the physiology of these animals, we cannot pronounce on the value of the mode of multiplication which the authors I have just cited propose to employ.

Whatever it be, after the entire results of which I render you an account, and after experiments analogous to those of Messrs. Géhin and Remy, made by M. Lefebvre, of Vaugorard, it seems clear that with perseverance, we can with little expense ameliorate the icthyological breed of France, and obtain also for our territory covered with water, a revenue much more considerable than that now derived.

This would be for the whole country, an increase of riches, and trials of this kind appear to me all the more important as several circumstances tend to diminish the alimentary resource of our rivers. The increasing rarity of fish, in a great number of our rivers, does not arise solely from the manner in which fishing has been pursued, but from other causes, among which is the extension of manufacturing industry. Thus the toll-gates established in such numbers for the service of hydraulic motors, are so many obstacles to the production of various fish, which require to ascend the rivers to their head-waters to find fit spawning spots, and single propagators arriving in smaller numbers in the streamlets, the fish interests of the river suffer, for the eggs are not in a condition favorable to the development of the young, and the means of recruiting the entire species is rapidly lessened.* If, as in Scotland, and even in England, there existed in France, many rich proprietors who possessed water-courses of considerable extent, we could leave to the care of private individuals all matters relating to improved river-fishing, for to whomever one of these streams belonged, he would be interested in increasing its products.

But with us it is altogether otherwise, and the individual who would occupy himself with stocking a stream with fish, could hardly hope to reap personal profit therefrom: he would augment the alimentary resources of his fellow-citizens, and thus

* It is worthy of remark that waters of paper manufactories, which contain so large a quantity of chlorine for whitening the rags, are injurious to fish. It is one cause of destruction worthy of note.

render his country solid service, but he alone would enjoy but a small interest in the benefit so diffused, and ordinarily would want the stimulus to undertake the labor.

The stocking of rivers, then, should be considered a work of public utility, and it seems to me that it is the business of the state to look after it.

Trials of this kind made on a great scale, and prudently conducted, and confided to intelligent men, would not involve heavy expenses to lead to important results. If you judge proper to have them executed, you will find in the two fishermen in question, capable agents, and I would add that the charge of such work would be the least recompense the government could make them.

For the rest, such an enterprise would necessitate serious preliminary studies, and give rise to several questions, for whose solution the opinion of the administration of waters and forests would be necessary, as well as the light of naturalists, and it would perhaps be necessary to have a mixed commission. To sum up —we perceive that the stocking of fresh waters with artificial methods was long since thought of, but it has only been tried in France lately ; that Messrs. Géhin and Remy appear to have been the first to put the method in practice among us, and that for their part they have arrived at results analogous to those obtained at the same period, in England, by Mr. Boccius; that the labors of these two fishermen are worthy of attention, and that in applying to the reproduction of salmon, the means they have successfully used to rear trout, we shall be enabled largely to increase the interests of our river-fisheries.

I have the honor, &c.,

MILNE-EDWARDS.

Report on the means of stocking all the streams of France with fish, addressed to the Minister of the Interior, of Agriculture and of Commerce.

PARIS, July 12, 1852.

SIR :—In your letter of the 30th June, you ask me to visit the fish-breeding establishment at Mulhouse, of Messrs. Berthol & Detzem, engineers of the Rhone and Rhine Canal, and to suggest to you measures so that their works can be made to stock all the streams of France. Accordingly, I now put you in possession of the result of this mission.

The discovery of artificial fish-breeding, was long hidden in the laboratories of science, where it remained confined to physiological experiment ; but lately it has been practically set forth by the Count de Goldstein, by Boccius, and above all, by the two fishermen of Bresse, and sober inquiry and trial have been adopted to attain to the precision of pure method in regard to it.

I have shown, for my part, with the assistance of Messrs. Berthol & Detzem, that not only the eggs of fish, brought from very distant waters, preserve all their native powers of conception, but that by means of machinery, extremely simple, they can be hatched much more quickly and certainly than as the female ordinarily lays them ; so that two sets are obtained in the ordinary time of one.

This double result, that of carrying without injury, eggs to a great distance, and their rapid fecundation, leads to the possibility of restocking all the streams of France, in a single season ; so that it will cost nothing to the state but the necessary advances to organize an establishment, wherein the spawn accumulated from all points where they are easily secured, should be confided to the care of the canal keepers. I say it will cost the state nothing, because the advances can be readily more than repaid by a contribution, voluntarily self-imposed by the proprietors in exchange for the precious gifts made them, whether in the form of eggs or young fish.

The more I reflect on the means of realizing this useful enterprise, the more I consider it our duty to insist that France shall take the lead in giving a practical example of this great scientific discovery, which can so increase public wealth by creating an inexhaustible means of production. It is a wish I express with all confidence, because I have visited the spots where the project has already received an impetus under the auspices of two engineers, who, notwithstanding their limited resources, have raised, this year, a million trout, salmon, and mongrels ; the greater portion of which they showed me scattered through the ponds which they have dug along the Rhone and Rhine canal.

It only remains to profit by the experience and devotion of which they have, during two years, given so many proofs, and to place in their hands sufficient means to transform the precarious arrangements due to their perseverance into a veritable establishment, where, as in the best regulated manufactories, the working details are ample and ready.

The locality which they have chosen, is admirably well adapted to their purpose : a stream of fresh water, clear as crystal, runs from the foot of a sheltering hillock on a common of several acres, and then branches off into smaller streams. This so well fitted to fish-hatching, especially of trout and salmon, could be easily turned into a vast breeding establishment. It would be only necessary to substitute for the sieve-boxes hitherto used (which offer obstructions and become less and less permeable), simple plates placed longitudinally in parallel partitions, which will divide the stream into narrow drains more or less numerous, through which the water will flow with some degree of rapidity. These drains intended to receive the eggs, will be cut at intervals so as to form a succession of falls, in order to hasten the course and give an airing to the water, and produce conditions most favorable to the end in view. Each one of these drains should be extended in a meadow, without being confounded with the others, and finish by enlargement in a spacious basin, where the water in question alone has access, and whither will come the young fish when hatched,—another place of destination being in store for them.

When this stream will have been so transformed into a vast establishment, made after the plan I have indicated, it should be covered with a glass roof like a greenhouse, admitting the light, and formed of movable panes turning round, so that the air may be readily admitted when deemed necessary. To this should be added a little house, to protect the workmen, where a workshop of all the necessary implements would be, and also a register of the results of each day's observations. The natural history of fish so obtained, would offer invaluable details. When this establishment would be ready, the problem would be reduced, simply procuring eggs sufficient to fill it, and thence to stock all the streams of France. This would not be difficult to realize.

Being on the frontiers of Germany, Messrs. Berthol & Detzem are in communication with the fishermen of the river and the great lakes where are fish the most esteemed. These fishermen have undertaken to give them all kinds of eggs. Messrs. Berthol & Detzem have already taken from Lake Federsee, thirty-six gigantic fish, which so transferred, I have seen in their basins. They are waiting now for a supply of young fish of this kind, which bear the journey so easily, that I obtained three for the College of France, by simply putting them under the care of the conductor of the diligence, who kept them two days and a night in a vase. These fish hatch even in turf-pits; so that they can be easily propagated in those of Picardy, and in the least favorable waters. Their importation, then, will be a service rendered to fish-breeding.

In hatching fish in new waters, trials of their acclimation can be successfully made. I may give here striking examples in citing my experiments at the College of France, under circumstances where I did not promise myself any success. Young salmon, hatched in my laboratory, and placed afterwards in an artificial pond fed by a single stream of the water of the Arcueil, grew as well as if they had lived in the Rhine, as I was able to satisfy myself by a comparison. They are hardly four months old, and already their length is 60 milemetres, of which they have gained 12 during the last twenty-four days; a remarkable growth, which may be attributed, without doubt, to the particular nourishment they receive, of which they show themselves greedy.

But to return to our hatching apparatus, and the eggs which are in progress of development:—Here a second problem is presented—what becomes after birth of the young fish hatched by millions in the narrow drains where the eggs were deposited? This second problem will not be more difficult to answer than the first. The arrangement of the locality will answer for all exigencies. As soon as the newly-hatched fish are strong enough to swim, they will follow the course of the stream, which will draw them to the meadow by the extremity of the glass-house through which the current passes, and leave them in the basin. There they will grow; but their number increasing every day, they cannot be long kept in this narrow reservoir. Larger basins then must be provided, where they can grow with proper nourishment. The dependencies of the Rhine and Rhone Canal will fulfil this office, and on a scale so vast, that there will be a crop greater than one would suppose room could there be found for. Thus:—The government has on the borders of the canal, on the right and left,—land in length 117,730 metres, and breadth 15 metres. Already there they have dug a certain number of ponds, well supplied with water. These ponds may be multiplied indefinitely and connected by gratings, so as to prevent the admixture of the different kinds of fish, and stopped off occasionally in order to admit of being severally emptied, so that the young fish can be taken from them. But the ponds already dug on one side of the canal are in the same part of the meadow with the receiving basins, into each of which the hatching trenches will carry a particular species; and it results from this, that to transfer the young of this species from the establishment where they were hatched to the ponds where they are to be converted into larger growths, there is almost nothing to do. The operation will be self-accomplished, so to speak; and from the single circumstance of a happy distribution of the different waters which run from one side to the other.

When the spawn have arrived at the growth of young fish suitable for stocking streams, the Rhone and Rhine canal, which runs between the two long lines of ponds where these fish are kept in reserve, will itself be the natural means to conduct them into all the waters of France by means of their intercommunications. To attain to this object, a jointed raft should be made of pieces of wood transversely placed, and connected by iron rings, and in the interstices of this raft would be fastened casks sufficient to hold the entire supply of fish. These casks should be provided with gratings so as to be permeable, and contain water-plants, so that the young fish are not injuriously crowded.

The convoy so disposed should stop successively before each pond, and right and left, the workmen attached to the ordinary service of the canal, will empty into it the fish drawn from these drains ; then the cargo completed, the raft will be set in motion, and the casks with their bottoms knocked out from time to time, will sow the fish, as a plow would sow seed if capable of doing thus as fast as it made furrows.

When the convoy will pass the point of junction of another water-course, one of its sections, as they are fastened by rings, could be detached as a wagon is from a train, and given to the engineers of the country traversed by this stream of water ; these engineers will take the portion of the convoy in question, in order to empty it in the localities which appear to them the fittest to the purpose, and so ascertained beforehand, and then will return it to the point of departure, so that on its arrival thither, the great convoy may unite all the detached fragments, and render them to the establishment in order to take a fresh load if the first has been insufficient, or to wait until a second crop requires a new journey.

The restocking of all the waters of France, will be accomplished then easily, since, on the one hand, the officers of the roads and bridges will answer for the requirements of the service, and on the other, the organization of the entire establishment will require but a first expenditure of 22,000 francs, necessary for the construction of the shed, the guard-house, the digging of the ponds, the purchase of tools, and of twenty acres of ground to be inclosed in the common already given by the municipal council of the locality.

The first expenditure, or an annual credit of 8,000 francs, will suffice to commence the work, to procure the species most valued, meet the cost of the daily labor, and give the production an infinite extension.

It will be perceived, therefore, that this sum is the smallest trifle compared with the riches it will produce, for here nothing less is aimed at than to keep the supply of food up to the increased consumption, according to the duty imposed on governments : hesitation in such case is allowable only when an adequate trial renders success doubtful; but here experience has already furnished such positive results, that there cannot be the least doubt of the success of the operation.

Time presses, sir; and there are only three months before we come to the breeding season of salmon and trout. If at that time the apparatus is wanting, we lose the most interesting part of the required work. I trust, then, you will give me the order for a credit of 30,000 francs, immediately open to the engineers of the Rhone and Rhine Canal, and I shall be happy to offer you my assistance for the organization of an establishment so founded, and to take my part in the responsibility of an enterprise which will be a signal honor to the administration.

I cannot terminate this Report, sir, without speaking to you of the propagation of fresh-water shell-fish; experiments which I have made under the hope of applying them to salt-water shell-fish, whose multiplication would not be difficult to secure. Here, then, is an account of these experiments:—I placed, at the College of France, in a basin, like that wherein my young salmon live, fed by a rivulet, a certain number of female craw-fish, all carrying under the tail their eggs. At the end of twenty-five days, all these eggs were hatched, and the basin was usurped by a myriad of young craw-fish, which grew perceptibly. This result proves how easy it is to restock all running streams which an abuse of fishing has devastated, as though they had never been supplied. The question is reduced simply to setting apart at the breeding season, in the reservoirs in the form of little brooks communicating with creeks or rivers, all the females who have their eggs attached to the appendices of the tail, and not to allow their consumption until their offspring is hatched. This offspring, retained afterwards for a period in propagating streams, would not be allowed to swim through the gratings until capable of taking care of themselves.

As to salt-water shell-fish, France possesses on the Mediterranean shore, immense salt marshes, where the females of these animals could also be retained till the moment of hatching their eggs, as they carry them under the tail like the craw-fish. If the experiment succeed, and these spawn increase on the spot sufficiently fast, they may be fattened in these vast receptacles. If, on the contrary, the conditions are unfavorable, they should be at liberty to go at large to seek another spot and stock our coasts.

But this is not the only use to which these marshes can be put. The sea-fish are too much liked not to suggest the means of multiplying them, either by artificial fecundation, or by transporting the young fish of certain kinds. In favoring the realization of such an enterprise, the state will have created in a few years, ponds much richer than the artificial piscines which were dug at so great an expense by the Romans, by the Gulf of Naples; piscines among which, however, those of Lucullus produced no less than four million sesterces, at a sale where presided Cato of Utica, in quality of tutor to the son of this famous epicurean. The care of these immense reservoirs would be confided to the customs officers of the coast, and would not involve, consequently, expense beyond that of fishing in the waters.

While these measures were taken to secure the multiplication of salt-water fish, it would naturally lead to the means of selling them for consumption at a price so moderate, that districts farthest off from their production could compete for having a supply of such alimentation for the laboring classes. You will find, sir, on this question materials for documents of great importance in practical details, from time immemorial, on the marshes of Commachio, whose waters are constantly changed by the flux and reflux of the Adriatic. There a population of about four hundred men, disciplined as if aboard ship, is occupied the year round, in fishing and preparing fish for all parts of Italy, with which they have a large commerce. It would be useful, then, to know the procedures by which they arrive at this last point.

Accept, Sir, the assurance of my most
distinguished consideration. COSTE.

LESSONS ON THE NATURAL HISTORY AND HABITS OF THE SALMON.

The following series of articles upon artificial and natural salmon breeding, appeared in *Bell's Life in London*, weekly newspaper, in January, February and March of the present year. They embody an account of what has been done in Great Britain, in relation to artifical breeding, and present some facts not found in the translations of the French works contained in this volume:

LESSON I.

HOW TO PRESERVE AND BREED IT ARTIFICIALLY.

At last the salmon is attracting, practically, public attention. Its present scarcity, compared with its past abundance, is the cause. If prevention had been practised—*obsta in principiis*—stop the evil in the beginning—it would not now be necessary to apply somewhat costly remedies. Happily real and effective measures are found, and all that is required is their general application according to proper formulæ.

The breeding of salmon, by artificial means, is now considered the last resource to replenish salmon rivers, formerly abounding in *salmo salar*, or true salmon—the genus salmo, the head of all the species and varieties of the salmon, known as *salmonidæ*, offshoots of the *fons et origo* of the race, and embracing every kind of trout from the *salmo ferox* to the smallest of rivulet trout, viz., the diminutive par. The artificial breeding of salmon has been taken up by the French Government, and placed under the *surveillance* of "*le Ministre de l'Intérieur, de l'agriculture, et du commerce,*" and under the practical application of M. Coste, "*Membre de l'Institut, Professeur au Collège de France,*" and of MM. Berthot and Detzen, "*ingénieurs du canal du Rhône au Rhin.*" The labours, and their results, of all these naturalists, together with those of our own, Messrs. Shaw, Andrew Young, Boccius, Milne-Edwards, are detailed in a work, entitled, *Instructions Pratiques sur la Pisciculture, suivies de Mémoires sur le même sujet*, by the Monsieur Coste already mentioned. As yet no encouragement has been given to the breeding of salmon artificially by the English Government within the British isles. Earl Grey did, and the Duke of Newcastle does, favor and support an attempt to transfer salmon to the rivers of Van Dieman's Land, by artificial means, under the direction of Mr. Gottlieb Boccius. One attempt has failed, Mr. Boccius says, through the retention, beyond the day fixed for sailing, by more than a month at Plymouth, I believe, of the ship, on board of which impregnated salmon ova were placed in tanks prepared with due care. This is not to occur in a second attempt, about to be made shortly. I have not any thing like implicit faith in the success of transplanting salmon from the rivers of this country to those of the antipodes, either by means of impregnated ova or living fish, young or adult. I repeat what I said once before, that if the rivers of Van Diemen's Land are to be stocked with salmon, it will be from impregnated ova or living fish, procured from the river Sacramento, in California. The transit from that country is by one-half shorter than it is from any of the British ports.

What the English Government has neglected to do, its subjects are now doing. An influential association for breeding salmon artificially has been formed in Scotland, at the head of which are the Duke of Athol and the Earl of Mansfield. They have begun, we believe, for now is the time, their operations in the Tay and its tributaries. The Messrs. Edmund and Thomas Ashworth, of Egerton Hall, Bolton, have purchased, in the Court for the sale of Encumbered Estates in Ireland, "A salmon fishery, extending from Lough Corrib to the sea," and have made experiments at Outerard, Co. Galway, in a report on which, signed W. H. Halliday, and dated Galway, 4th July, 1853, it is considered that there were "40,000 ova (impregnated) deposited; and, assuming that one third may not have come to matury, we may conclude that we have upwards of 20,000 young salmon (salmon fry) now living in these ponds, beyond the reach of their natural enemies." We shall see next summer and autumn how many of these young salmon will return from the sea into the river or rivers into which they shall have been put, in the fry or smolt state—how many shall return grilse of the average weight of 5 lb. Then the success of the artificial breeding of salmon on Messrs. Ashworth's plan will be tested. In addition to the above experimentalists we have Mr. Isaac Fisher, banker, of Richmond, Yorkshire, associated with other gentlemen of that town and county, breeding salmon artificially in the river Swale. In a letter recently written to me by Mr. Fisher, I find the following paragraph:—"To-morrow I am off to the Wear, where there is an obstruction to the ascent of salmon up our river, the Swale, and I hope we shall be able to carry our work so as to overcome this obstacle. We intend to prosecute our experiments on artificial breeding this winter with greater care, and more extensively than we have yet attempted. I am going over the acts of Parliament touching salmon fisheries now in force, and my friend, Mr. Thirwall, is also at work. I have named the object (the submitting to the sanction of the Legislature a salmon extension act) we have in view to many gentlemen in this quarter. I have promises of support to a great extent. Both our members will assist in carrying any bill that we may bring forward, that is, of course, if it be one that shall be carefully drawn up, &c.; and I really think if all set to with a will, we shall be successful. In France, England, Scotland, and Ireland, efforts are now being made almost simultaneously to propagate salmon by artificial breeding. The success, or otherwise, of these efforts in the British isles, will be known next season by the results of the Messrs. Ashworth's experiments in Ireland, and the season following by those of the Scottish association, supported by the Duke of Athol and the Earl of Mansfield, and also by that in Yorkshire, promoted by Mr. Isaac Fisher and coadjutors.

The question of breeding salmon artificially is not new to this country. The first British artificial breeders of salmon are Mr. John Shaw, of Drumlanrig, and Mr. Andrew Young, of Invershin, Sutherlandshire. The French and other artificial breeders have followed in their wake, I cannot say whether advantageously or not. I have my doubts. The lapse of a year and a half will (*si vixerim*) set them at rest. Mr. Shaw began his experiments in artificial breeding in January, 1836, and in 1840 published his *Experimental Observations on the Development and Growth of Salmon Fry, from the Exclusion of the Ova to the Age of Two Years.* Mr. Young began artificial breeding in 1841, chiefly to prove that young salmon became smolts in the twelfth month of their existence, then migrated, and returned to their native rivers the same year as grilse or young salmon, coming back to breed for the first time. Mr. Shaw, in his *Experimental Observations*, maintained that salmon fry did not attain the smolt state before a period of two years passed in fresh water, and did not migrate to sea earlier. He also maintained that parr and salmon fry were identical. Mr. Young maintained the contrary, and I believe that Mr. Shaw now acknowledges that he was in error, and that Mr. Young was right. I have never met with a salmon fisher, learned in the natural history of salmon, who did not agree with Mr. Young. The tardiness in the development of Mr. Shaw's artificially-bred salmon fry arose, in my opinion, from his having taken the impregnated salmon ova from the river Nith, and placed them for incubation in ponds fed by mountain rills. The artificial spawning-beds of Mr. Young were made in the river Shin, and the impregnated ova taken *out* of it. In 1848, Mr. Young wrote in the *John O'Groat's Journal*, published at Wick, a series of essays on the salmon, which I transferred the same year to the columns of *Bell's Life*. Mr. Young soon after collected them, and they appeared in book shape, under the title of *The Natural History and Habits of the Salmon, &c.* In that pamphlet he sketched the mode of breeding salmon artificially. In 1850 I wrote the *Book of the Salmon*. It is divided into two parts,—the first, by myself, is divided into four chapters, and treats of "theory, principles, and practice of fly-fishing for salmon, with lists of salmon flies for every good river in the empire." The second part contains two long chapters on "the natural history of the salmon, all its known habits described, and on the best way of artificially breeding it." The information embodied in these two chapters was obtained by me from Mr. Young orally, and from his published and private writings. I believe in the correctness of that information, and I believe it—illustrated as it is with colored plates after nature of the salmon—*ab ovo* to the smolt state inclusive—the most valuable as yet published on the history and habits of the salmon, and on the means of breeding it artificially. In a letter from Mr. Young to me, dated December 17, 1849, and part published in the *Book of Salmon*, pp. 158, 159, and 160, he writes, "I have been experimenting on salmon for upwards of 30 years. Few, I believe, if any, have paid the attention to the habits of salmon I have done. To enumerate all the experiments I have made would fill volumes. On this point I must abridge. In 1834, and for a number of years following, we [I suppose Mr. Young and his assistants] marked spawned fish for the purpose of settling the question, denied by many, of the return of salmon to their native rivers. This we did satisfactorily. We, in 1835, marked smolts to ascertain and set at rest the following point, denied by many, viz., that the smolts returned grilse the same year they first went from the rivers to the sea in the smolt state. The experiments proved this also; and specimens of the grilse that we marked when smolts, and which returned grilse from the sea to fresh water the year they were marked, may be now seen in the Museum of the Royal Society of Edinburgh. We continued these markings many years, invariably with similar results, and at the same period, continuously during three years, we carefully watched the spawning operations, and spawning beds in all their stages, and were fully convinced at last the fry remained in the rivers *one whole year, and no longer, after having been hatched.* However, though I was fully convinced in the case, the public were not, and still hung to the old theory that they were fry and smolts the same year [the fry produced in 1854 from ova deposited this year, say now, will not become smolts until the spring of 1855], and that their migration to the sea took place shortly after they were hatched. To make assurance doubly sure, we, in 1841, erected a chain of four artificial breeding ponds by the river Shin [I have seen the remains of them. They were in the Shin, close by its left-hand bank, and about a quarter of a mile or more from its mouth], about fifty yards above Shin bridge, when we hatched the fry to the state you have seen them preserved here. [In Mr. Young's museum at Invershin.] We continued this process for some years, and always found the same result. The ponds were visited and examined by Dr. Travers Twiss, of the University of Oxford, through whom I presented, in June, 1834, a set of ova and fry up to the smolt state." Who can doubt after this that the silvery-coated smolt, in length and weight about the admeasurement of a very large sprat, is a young salmon about a year old, and that at that age it migrates for the first time to sea, and does not wait until it is two years old to assume the silvery and migratory coat, as Mr. Shaw maintained it did. By Mr. Young's discovery we have a very important fact proved, viz., that grilse, weighing from four to eight pounds, are young salmon of 15 or 18 months old, a little more or less; that they breed towards the end of their second year, and that they are adult salmon in the middle, and not unfrequently in the early part of the third, year of their existence.

Before I proceed to analyze M. Coste's *Instructions Pratiques sur la Pisciculture*, &c., and *A Treatise on the Propagation of Salmon and other Fish*, by Edmund and Thomas Ashworth, the latter being little more than a translation of the best parts of the former work, I shall lay down briefly a few salient items of the salmon's natural history. In order to preserve that valuable fish, and to multiply it by artificial breeding, its history and habits, as far as they have been discovered, should be known.

The salmon is a fresh-water fish. In fresh water it breeds, and remains in it during the whole of the first year of its life. As long as it lives it passes, on an average, two thirds of every year in fresh water.

It never breeds in lochs, lakes, pools, or deep still water, but invariably in fords and shallows, and always returns after its annual, sometimes bi-annual migrations to sea, to the rivers in which it was bred, provided it escape during its immigrating voyage through salt water, destruction by fish of prey, amphibious animals, or by the devices of man.

It does not return from the sea to fresh water for the proximate or immediate purpose of spawning. If it did, we should not have fresh-run fish in January, February, and March. Few salmon breed before October. The general breeding time is the latter end of November and the beginning of December. Salmon emigrating from the sea in the first months of the year will occasionally make a second sea voyage in the summer, and return in the autumn to their native rivers. When salmon are surfeited with sea-found food, and have become full-fed and fat, they grow tired of salt water and its feeding grounds, and make for the estuaries and rivers. If the salt or brackish waters of the former are not well tinged with fresh water, the salmon remains in them until there be a flood in the rivers. They know when this takes place by the increased quantity of fresh water rushing into the narrow firths. Fresh water being lighter than salt water, it flows above the latter, and up into it the salmon swim, gambolling and swimming rapidly for the swollen river, its state enabling them to surmount weirs, cruives, and other obstacles, which would obstruct their passage if the river was low. In dry summer weather very few salmon are found in the rivers. A flood comes, subsides in two, three, or four days, and then the rivers—I mean good ones—abound with fresh-run fish.

A fresh-run fish is known by the brightness of its scales, by its corpulency, the whiteness and softness of its fins, by the shallow forking of the tail, and by parasitical insects adhering to it. These insects, or vermin, disappear, it is stated, from the sides and belly of the salmon after it has been in fresh water forty-eight hours. Some persons argue that salmon return to fresh water to get rid of these sea parasites. Not so ; if they did, they would, as soon as the insects disappeared, emigrate again to sea. This they never do until they have been many weeks in fresh water.

Salmon, after its first year, never grows in length or bulk in fresh water. After its first immigration from the sea as a grilse, it diminishes in muscle, fibre, and fat, every day it remains in fresh water. Its fins become black and strongly elastic, and its gill-covers and back assume the same color. It is then called a "black" fish, in contradistinction to the bright, clean, fresh-run (just arrived from sea) fish. It does not, until within a very few weeks of spawning, lose strength in fresh water. It is more active than the fresh-run fish, has greater propelling power by means of fins hardened by fresh water. Salt water so softens the fins of salmon as to render them feeble propellers. We do not know how far seawards salmon travel to feed. I should say not far beyond the mouths of estuaries or the shores of ocean. I do not think the salmon a deep-sea fish. If a river runs into a sea or frith, having a northern and southern direction, salmon, when they emigrate to either, take towards the north in search of feeding grounds, and return southwards to their native rivers. When the salmon resort to a sea lying east and west—the English Channel, for instance—I do not know which direction they take—whether they go eastward or westward. Would any naturalist of Hampshire or Devonshire tell whether the salmon of the rivers of those counties, when repairing to their salt-water feeding-grounds, proceed in the direction of the Atlantic or of the German Ocean?

This lesson is now long enough. I have not mentioned some of the minor habits of the salmon. I shall do so hereafter, not forgetting some of its wonderful instincts, great and rapid powers of digestion, and consequent incredible voracity, and rapidity of increase. I hope that those who are engaged in breeding the salmon artificially, who have studied or are studying its history and habits, will assist me during the winter months in the objects I have in view—preserving and multiplying salmon so as to render it as abundant and cheap in our markets as the cod-fish. They can assist me by correcting any errors into which I may have fallen, or by giving additional information on points upon which I have not been sufficiently minute, or by dilating on points that have escaped my notice, or of which I am ignorant.

LESSON II.

SALMON anglers, salmon lovers, salmon eaters, sing, "Oh, be joyful," and be so. More than "looming in the future" is a good time for you. I know not looming;—that won't do for me. I want the positive and palpable, and must and shall have it. I want salmon abundant and cheap, and in two years —no looming here—I, and you, and every body will have it. I mean us, poor people, for when the salmon is cheap, and is found as good and as plentiful near the Brill, Somers Town, as in Bond-street, Charing-cross, or Cheapside, your courtier will have none of it. Once become the food of the people, 'twill stink in his nostrils, and, pah, he'll look upon it as "a slovenly corpse." *Ainsi soit-il*—Amen; *per omnia secula seculorum.* We have begun the new year well. Read again and again "Salmo's," and "Y.'s," "Piscator's," and "Outis's" letters, in our impression of the 1st of January, 1854. In them behold the seed that will bring forth—in them see the forerunner of other sowers—in them contemplate the nucleus of a great and triumphant salmon-league —in them imagine the vendors and the makers of vendors of salmon selling it at 3d. per lb. It is pleasant to see three writers, one living in the Scottish Highlands, far, far away, another in Buckinghamshire, and the third, the erudite, acute, and correctly-calculating "Salmo," in Lancashire, totally unknown to each other, uttering identical opinions—at least upon one great point—touching the salmon question. It is pleasant to me to see such men responding to my call for aid, and agreeing with me on most points. If we differ—if I dispute a point with them, it is in the most friendly manner, for the good of the cause so dear to us all, and for nothing in the whole world besides. Last Sunday "Salmo" wrote, "My experiments on salmon were discontinued last year, as I had not an opportunity of visiting the Hodder at the proper time; they have, however, been resumed this winter under very favorable auspices. In the mean time, I have continued my experiments with trout successfully, and I shall give you the result in some future paper. Before entering upon that question, however, I purpose saying a little more about the breeding and preservation of salmon." I beg of so able a writer not to forget his promise, and I also beg him to bear in mind the vast circulation of this paper amongst sportsmen and naturalists. It is read, I am certain, by half a million of persons, so that "Salmo" will have a very large auditory. The larger it is, the more good he will effect. Both "Salmo" and "Y." are of opinion that artificial breeding is necessary only in salmonless rivers. In fact, they think it injurious in rivers in which there is a fair amount of salmon. The following calculation by "Salmo" is curious. He says, "Assume that a full-grown salmon contains 10,000 ova, which is considerably under the mark. The salmon choose for their spawning places, technically called 'Ridds,' the rough beds of gravel which connect the foot of one pool with the head of the next. In the Hodder these gravel beds occur at intervals, varying from 100 to 300 yards. Assume that there are five such spawning places in one mile of water, then ten miles of river would contain fifty of them. Assume that 10 fish spawn annually on each of these gravel beds, the result would be, that, in these ten miles of water, 500 salmon would produce 5,000,000 of ova, which, if they arrived at the maturity of their parents, would extend in a continuous line, head to tail, upwards of 2,000 miles. Now these results are so startling, as to prove at once that it is not from the deficiency of young fry that we have to lament the decrease in the number of salmon. The mischief must be sought for elsewhere. If only one fish in a hundred of those which are bred in the river returned to it mature salmon, we should have to boast of 50,000 annually in the Hodder; yet, for the purposes of this illustration, I have assumed only 500 pair of breeding fish." No doubt salmon fry are not sufficiently protected, and no doubt that if "one in a hundred" returned to its native river, we should have abundance of salmon. But I conceive this is no argument against artificial breeding. Will artificial breeding increase or not the number of salmon fry? If it do, though the loss will be greater, the return of them as grilse will also be greater. In well-preserved rivers, such as the Shin in Sutherlandshire, or the Erne in Ireland, in which the fishings cease in the third week of August, artificial breeding may not be requisite, but in rivers badly cared for, or over-fished, I think it may be profitably brought into requisition. My excellent friend and old correspondent, "Y." writes, "Although we allow that artificial breeding *may* be a good supplementary fund to natural breeding, the amount of that benefit is yet wholly to be seen. It is all on paper, and not yet in reality. Artificial breeding is valuable for the stocking of barren rivers, and the discovery for that purpose is of the utmost importance. But the attempts now in progress to increase the numbers in almost fishless rivers that negligence and bad laws have produced, are only yet in embryo, and not too much dependence is to be placed on them—at all events, not that dependence that should induce us to underrate that beautiful production of fishes which nature has so distinctly granted them. We are aware that long previous to the time that Mr. Jacobi wrote an account of his artificial propagation of fishes in the *Hanover Magazine*, that salmon, in Britain and Ireland, flourished and increased to an incalculable degree, under just and natural laws, and by natural breeding, and I have no hesitation in saying that they would do so still. The proprietors on Tay for the last two years acted wisely towards that river, for they trampled the present base act under foot, and closed the fishings on that river on the 26th of August, in place of the 14th of September, and by that means they have got the river better supplied with breeders than it has been during the existence of the present laws." There is much caution in these remarks. It is evident, however, that "Y." is not favorable to artificial breeding except in "barren" rivers. He does not speak out so boldly as "Salmo," but I feel assured that on this point their opinions tally. "Y." speaks well of the condition of the river Tay. It is surprising, then, that some of its proprietors should have recourse to artificial breeding, or, if not surprising, it must be taken as a proof that they hope some good at least from breeding artificially. And here, *en passant*, I will hint that I do not much like the breeding-boxes of Mr. Ramsbottom used by him upon Tay and in Galway. Why not breed in the bed of the river withinside a longitudinal dam? Of this I shall have to say a good deal hereafter. If I were convinced of the accuracy of the following calculation set forth by "Salmo," I should be obstinately opposed to breeding salmon by artificial means. He says, "It must be remarked, that only a limited portion of the ova of a salmon is mature at a time; that to obtain the gross produce of one fish, you must handle, at least, 10 or 12, probably 40 or 50 fish; that, in procuring these fish, you must disturb a considerable number of others, and interrupt their spawning at a very critical period; that every fish so handled, is more or less injured, and rendered for the time quite sickly and helpless ; and that, do all you may, your utmost exertions will not enable you to collect so much as one per cent. of the ova deposited in the river ; and that, after all, you have only robbed the river of so many ova which would otherwise have been deposited there without your aid." Now, in my opinion, if the female salmon be properly manipulated the ripe ova will be expelled only, and if the fish be returned to the river, she will recover, her remaining ova will become mature, and she will deposit them, and get them impregnated as if nothing had happened. I believe this to be Mr. Andrew Young's opinion. At the best, however, it can only be opinion or surmise. It is very difficult to tell what becomes of a salmon returned to the river after some of its ova have been expressed for the purpose of producing salmon in boxes or ponds. A late number of the *Perth Courier* says, "On Saturday last, from five female and one male salmon, caught below a ford near the mouth of the Almond, about 50,000 eggs were got." This statement does not agree with "Salmo's" calculation.

I have as yet said little of "Piscator's" letter. He clinches the matter at once by saying that the only remedy to prevent the decay of salmon is a change in the salmon-fishery laws—a change from bad to good. It is the best remedy, and, if all of us were like "Piscator," we soon should get it ; if all were as ready as he is to form "An Association for the Protection of Salmon," the Legislature would accord the remedy required. "Salmo" mentions the large amount of fry and smolts to be seen in even tolerably good salmon rivers, and how few of them return as grilse. How can they? It is not generally made penal to kill salmon fry. I know of no trial for the offence, except one that took place last year before a meeting of magistrates at Worcester. Yet not far from that town they are captured wholesale, and served up at the inns as a delicious *plat*, under the name of "lastsprings." In the Monmouthshire Wye they are caught by bushels, and so they are in every Irish river with which I am acquainted. I have seen urchins on the Suir, county Waterford, and in the Lee, county Cork, catch them by scores with the artificial fly, which they take with extreme avidity. Hundreds of thousands of them are destroyed by eel-traps and at mill-dams, and there is at present no help for it. I believe that many persons are not aware that salmon fry is young salmon. They think them trout, and, whilst slaying them, are not aware of the mischief they are doing. The Reverend author of the "Erne; its Legends and its Fisheries," mentions in that interesting work that he used to see, near Ballyshannon, hundreds of boys "whipping" for salmon fry, and no one durst interfere with them for fear of parental vengeance. All this must be stopped. The coming season will be an abundant salmon one; and the following better. We are now beginning a real salmon agitation; and

"I think I hear a little bird that sings,
The *salmon* will be masters by and by";—

Legitimately masters, pronounced so at the bar of the House of Lords by the binding words, "*La Reine le veut.*"

In my first lesson of the date of Dec. 8, I gave a few items of the natural history of salmon. I then promised that others should in due time follow. I now redeem my pledge. I know of no creature whose growth is so rapid as that of the salmon. The smolt weighing between two and three ounces, becomes in three months a grilse of six pounds, more or less. Mr. Young has seen a grilse that weighed fourteen pounds, and one that weighed little more than one pound. The largest and smallest, I fancy, he ever saw during his long experience of 40 years and more. The large grilse must have remained at sea an unusual length of time, and have been descended from Patagonian parents; the lesser one must have only remained on the salt-water feeding grounds three or four weeks. In *The Book of the Salmon*, pp. 199–200, I have written, "This growth of salmon at sea, and at sea only, after having obtained in fresh water the smolt size, depends on three things: duration of time they remain on their sea feeding-grounds, quality and quantity of food they obtain thereon, and hereditary capacity for growth with apportion-capacity for digestion. The grilse of small salmon, that is, of salmon which never grow beyond a small size, are handsomer, in every way better shaped, and generally of a brighter silvery hue, than the grilse, the produce of larger growing salmon. The grilse of the rivers Carron and Laxford, in Rossshire and Sutherlandshire, are handsome, small-headed, thick and deep, and short in the body ; their scales are small, smooth and bright, and all this because they are the offspring of small, well-shapen parent salmon : whereas the grilse of the river Shin, in which salmon grow to a very large size, are ill-shaped fish, having large heads, long, thin bodies, large long fins, and large, rough, and by no means brilliant scales. It requires experience to distinguish a large and well shaped grilse from a small salmon. Very frequently the only distinguishing marks between grilse and salmon are the smaller scales of the former, the longer and larger fins, and the more forked tail." The powers of digestion of the salmon are amazing. I opine that a salmon would digest a herring in a few minutes, and I think that whilst at sea the quantity of fish-food it consumes is immense. Hence the chief cause of the wonderful rapidity of its growth. A salmon never grows so fast as in the second year of its existence. An adult salmon, weighing 10lb, has been known to more than double its weight in 37 days !

This ratio of increase cannot continue long, and I conclude that very large and aged salmon remain stationary as to growth. The male salmon at spawning time, and in its kelt state, has a soft tooth-like excrescence in the lower jaw fitting into the upper.

The use of this excrescence is not known. Salmon in surmounting weirs and waterfalls jump straight upwards—not perpendicularly, but rising gradually as a man taking a running leap over a hurdle, hedge, or gate. On the moot point of salmon leaping, I say, in a *Handbook of Angling*, "Natural historians used to gravely tell us that salmon, in order to jump high, were in the habit of placing their tails in their mouths, and then bending themselves like a bow, bound out of the water to a considerable distance, from twelve to twenty feet." The late Mr. Scrope (*Days and Nights of Salmon Fishing*), correctly culates that six feet in height is more than the average spring of salmon, though he conceives that very large fish, in deep water, could leap *much* (which I doubt) higher. He says, "Large fish can spring much higher than small ones; but their powers are limited or augmented according to the depth of water they spring from: in shallow water they have little power of ascension; in deep they have the most considerable. They rise very rapidly from the very bottom to the surface of the water by means of rowing and sculling, as it were, with their fins and tail; and this powerful impetus bears them upwards in the air, on the same principle that a few tugs of the oar make a boat shoot onwards after one has ceased to row." The ascending motion is caused by the salmon striking the water downwards with its pectoral, ventral, and dorsal fins, aided by bodily muscular action.

Salmon invariably spawn on gravelly and sandy shallows. As soon as they have paired—the females seeking the males, as it is said maids do of leap-years—they choose a fit spawning locality, from which, if they can, they chase away all other fish. For some days they are engaged in this operation. The coast being clear, and the female ready to lay her mature ova or eggs, they commence constructing what I call their nests. How these are situated and constructed I explain, on the *viva voce* authority of Mr. Andrew Young, in *The book of the Salmon*, p. 174 *et infra* :—"The spawning-bed, which may be called a continuation of nests, is never fashioned transversely, or across the water-current, but straight against it. The way the bed is formed has never before been accurately described. Some have affirmed that the male fish is the sole architect; others, that the female does all the work; others, again, that the tail is the only delving implement employed; and others write that the bed trenches are dug across the stream. A salmon spawning-bed is constructed thus :—The fish having paired, chosen their spot for bed-making, and being ready to lie-in, they drop down stream a little, and then rushing back with velocity towards the spot selected, they dart their heads into the gravel, burrowing with their snouts into it. This burrowing action, assisted by the powers of the fins, is performed with great force, and the water's current aiding, the upper part or roof of the excavation is removed. The burrowing process is continued until a first nest is dug sufficiently capacious for a first deposition of ova. Then the female enters this first hollowed link of the bed, and deposits therein a portion of her ova. That done, she retires down stream, and the male instantly takes her place, and, pouring, by emission, a certain quantity of milt over the deposited ova, impregnates them. After this the fish commence a second excavation, immediately above the first, and in a straight line with it. In making the excavations they relieve one another. When one fish grows tired of its work it drops down stream until it is refreshed, and then, with renovated powers, resumes its labors, relieving at the same time its partner. The partner acts in the same spirit, and so their labor progresses by alternate exertion. The second bed completed, the female enters it as she did the first, again depositing a portion of ova, and drops a little down stream. The male forthwith enters the excavation, and impregnates the ova in it. The different nests are not made on the same day, but on different days, progressively. It is never so all at once. The ova in the first nest are covered with gravel and sand, dug from the second, being carried into it chiefly by the action of the current. The excavating process just described is day by day continued, until the female has no more ova to deposit. The last deposition of ova is covered in by the action of the fish and water, breaking down some of the gravel crust above and over the nest. Thus is formed a complete spawning bed—not at once, not by a single effort, but piecemeal, and at several intervals of greater or less duration, according to the age and size of the fish, and quantity of ova to be deposited. A female salmon in its third year has a larger quantity of ova to deposit than a female grilse, or young salmon in its second year ; and it may be taken for granted that the older and larger either fish, male or female, is, the greater the quantity of ova to be deposited and of milt to be emitted. In consequence, the time occupied in deposition chiefly depends upon the size and fecundity of the female fish. The average time is from five to ten days. It would be more correct to say the mean time lies betwixt." As soon as salmon have spawned they are kelts and foul fish, totally out of condition and unfit for human food. They drop down into the pool next below the spawning bed, and there remain until they have somewhat recovered from the exhausting process of procreation. They then proceed slowly seawards, and by the time they approach the mouths of rivers they become "mended" kelts. In this state they eagerly take the artificial fly and other baits, but though the angler, fishing for fresh-run fish, cannot help capturing them, none but the arrant poacher will keep them in captivity. The true sportsman will take them tenderly off his hook, replace them in the river, that they may go to sea, and there grow and fatten, and come back a clean, beautiful fish, in high condition.

I have now touched on the salient and important points of the salmon's history and habits. There remain still some curious and amusing ones to be mentioned. We defer them to another day. My third lesson shall be on the methods of breeding salmon artificially—on M. Coste's, member of the French Institute and professor at the College of France; on Mr. Young's, as given in the *Book of the Salmon;* and on that of the ingenious and indefatigable Mr. Gottleib Boccius, should I see him within the next week or so.

Jan. 4, 1854. EPHEMERA.

LESSON III.

As I am not so much opposed to the artificial breeding of salmon as several celebrated corespondents of this journal, I shall briefly sketch the different methods formerly and recently adopted in Europe for propagating salmon by what are called artificial means. The Messrs. Edmund and Thomas Ashworth, of Egerton Hall, Bolton, Lancashire, have purchased in the Court for the Sale of Encumbered Estates a "several fishery, extending from Lough Corrib to the sea," and have attempted in the spawning season of 1852 to propagate salmon near Outerard artificially. Last year, they published "A Treatise on the Propagation of Salmon and Other Fish." It is not original, being merely a translation of certain portions of M. Coste's French work on pisciculture, extracts from Jacobi, and Messrs. Young and Shaw, concluding with Mr. Halliday's report of the experiments made at Outerard. Nevertheless, though not original, it is a useful pamphlet. There is a plate, copied from M. Coste, containing diagrams of the utensils used in artificial spawning, with figures of salmon from a day to ten months old inclusive.

The first discoverer in Europe of artificial spawning was Jacobi, a German naturalist. In 1773, exactly ten years after Jacobi had developed his theory in the *Journal of Hanover*, it was translated into French by Duhamel du Monceau. In this country Mr. Shaw of Drumlanrig began to breed salmon artificially in the year 1836, and Mr. Young of Invershin, a few years later, viz., in 1841. The Scotch breeders succeeded in producing salmon artificially, but they differed widely as to the ratio of growth of salmon so produced. Mr. Shaw maintained that salmon-fry did not attain the smolt or migratory state until it was two years old. Mr. Young contended that it did at the completion of its first year. Subsequent observations and experiments have proved Mr. Young right and Mr. Shaw wrong. Messrs. Shaw and Young are, therefore, the first artificial breeders of salmon in Great Britain. In France, in 1851, the system was first publicly adopted by MM. Berthot and Detzem, at Huningen, near the Rhine. They were preceded by private experiments made by two fishermen of Bresse, MM. Gehin and Remy.

M. Coste, member of the Institute, and professor at the College of France, next took the subject up, and supported by the French Government, he has been, and is carrying on in France the artificial propagation of salmon, trout, and other fresh-water fish on a very large scale. We do not know as yet fully the results. All we as yet know is, that the French breeders have produced large numbers of fry, but we do not know how many of them have arrived at maturity—how many have attained marketable value.

I shall pass over Jacobi's method of artificial spawning, and come to that carried into effect at the establishment of the Messrs. Ashworth, at Outerard. It is founded on M. Coste's plan. Mr. Halliday repórting on the spawning establishment at Outerard, writes from Galway, July 4, 1853—"Robert Ramsbottom, from Clitheroe, was sent over by Messrs. Ashworth. The plan tried was by spawn boxes, prepared, and by an artificial rill-bed, running parallel, and both were equally successful. On the 14th December, 1852, a small rill at Outerard was selected for the experiment ; by a rude check thrown across, a foot of water head was raised over a few square yards, to insure regularity in the supply. From this head, half-foot under surface level, three wooden pipes, two inches square, by a few feet long, drew off respectively to the rill-bed and to the boxes all the water required—the surplus of the supplying rill passing away in its usual course. The boxes are six feet long, eighteen inches wide, nine inches deep, open at top, set in the ground in a double row, on a slope of two or three inches on each box, the end of the one set close to the end of the other, in continuous line, and earthed up to within an inch of the top. They are partly filled—first, with a layer of fine gravel, next coarser, and lastly, with stones, coarser somewhat than road metal, to a total depth of six inches. A piece of twelve inches wide by two inches deep is cut from the end of each box, and a water-way of tin nailed over this, with a turn-up on either side, to prevent the water from escaping. These connect the line of boxes, and carry the water to the extreme end, whence it is made to drop into the one which receives and preserves the young fish. The artificial rill is, in all respects, similarly prepared, excepting that its channel course is in the rill itself. The pipe is now introduced into the upper box of each line, and at the water head the spawn bed is prepared, two hours' running will clear away the earth from the stones. The water will be found about two inches in depth, over the average level of the stones in the boxes. By an iron wire grating, the boxes can be isolated, and the pipe protected against the passage of insects and trout. The salmon were taken by nets on the spawn fords at night, from the 20th of December, 1852, till after the 1st of January. When taken, they were instantly, and without injury, put into a tub, one-fourth full of water.

The female fish was turned over, one man holding the tail, another running his hands down each side from the head, and, pressing lightly with his thumbs, the ova were readily discharged into the tub; a similar course readily discharged the milt. Both fish immediately, and apparently without the smallest injury, were returned to the river. The contents of the tub were then mixed by a motion of the hand. In one minute the water was poured off, and fresh put on, which was also poured off, and the ova put into the vessel, to carry to the prepared hatching ground. In pouring off the water from the ova, always retain sufficient to preserve it from the air, both in the carrying vessels and spawning tubs; unless the fish be in a fit state the ova will not shed by gentle pressure, in which case no violence should be used, but the salmon returned to the river, and fish in a more advanced stage taken. In distribution, the ova intended for one box should be put into one vessel, and this poured out gradually at the upper end of the box; the waterflow downwards will carry it [them, the ova] among the stones, under which it [they] will settle down, and wherever too thick, by raising some water in the vessel, and pouring it down, this will disturb and float the spawn into a more equal distribution, that should, if possible, be done the same night as taken. We consider the boxes used of sufficient size for 3,000 ova each; and, as a guide to the quantity found, an English half-pint will contain about 1,200 in number. We consider there were 40,000 deposited, and, assuming that one third may not come to maturity, we may conclude that we have upwards of 20,000 young salmon living in those ponds, beyond the reach of their natural enemies. From the lateness of the season, although the numbers of spawn fish lifted were very considerable, the above quantity of ova might be readily obtained out of five pairs of full, good brood fish, and that a million of ova [more correct to say 800,000] might, by a similar process, be deposited from one hundred pairs of salmon. A very curious fact was also ascertained in the course of this experiment. In taking up the spawning salmon we also caught a quantity of trout; those we examined, and, *in every instance save one*, they contained salmon ova, on which they were preying. From the gullet of one large trout we estimated that 600 were by pressure ejected, and I retained them along with a further quantity from other trout, and deposited all in boxes, isolated from the others, a considerable portion of which came to life, and are with the other fry in the ponds at Outerard, where thousands of young salmon may now be seen. And this experiment again shows that the year of their deposit as ova is not that of their migration to the ocean; until that period, it is of great importance to retain the young fry in these ponds, where they are protected from their numerous natural enemies; hereafter they must necessarily be left to protect themselves, and are more capable of doing so ; up to that time they should be occasionally fed with suitable food, and all the fish in one pond should be of similar age, the larger fish proving very injurious to the smaller and weaker ones."

The spawn deposited in the Outerard boxes in the middle of December, 1852, would be hatched about the middle of April, 1853, and, therefore, will be twelve months old next April, and those deposited 1st January, 1853, will be about twelve months of age in May next. According to Mr. Young's theory, and mine, they will then be silvery-coated smolts, and will about the time migrate to sea. Will Mr. Halliday mark a few score of them when they assume their migratory coat, and so ascertain how many will return from the sea as grilse, and what will then be their average size? If I understand right, all the salmon fry will be kept in the ponds until their period of migration. I regret this, for if a portion of them were freed from the ponds, and let into the river at two or three months old, and marked, an important question would be decided, viz., whether it were better to confine young salmon in ponds until the smolt state, or turn them when very young fry into the river from which the ova were taken. If a greater number of pond-imprisoned smolts, than liberated fry—the number imprisoned and liberated being equal—returned from sea to the parental river as grilse, then pond confinement would be proved more beneficial than early liberation. At present I do not think so. Non-migratory fish, the common trout, for instance, may be beneficially kept and fed in artificial ponds or streams for a year. I am not at all sure of the good expected to be the result of so confining migratory fish. I suspect they may never return to the rivers into which they shall be turned from the rearing ponds. If Mr. Halliday will adopt my suggestion of marking his smolts, he will dissipate my doubts and those of many others interested in every thing that pertains to the natural history of the salmon, in its increased propagation and preservation.

This lesson would have contained other modes of breeding salmon artificially had not a flood of sporting *news* set in, overflowing our wide margins, and leaving little room for sporting *essays*. I shall, as soon as the flood subsides, resume, and, I hope, conclude, the questions of salmon breeding, rearing, and preserving

EPHEMERA.

Jan. 20.

LESSON IV.

I HAVE received such powerful aid from those clever, conscientious, and practical correspondents, "Salmo" and "Y." that nearly all the task I proposed to do myself has been executed by them. I could not have done so well as they have. To their communications, which appear in company with this, I direct the careful attention of naturalists, and of all who own salmon rivers, and are desirous of having productive salmon fisheries. Printed proofs of them, and of my lessons, shall in a few days be forwarded for the consideration of H. R. H. Prince Albert, who has always exhibited beneficent anxiety and readiness to add to the comforts of the English people, to the Duke of Newcastle, Lord Palmerston, Mr. Sydney Herbert, and Sir W. Molesworth, all of whom are desirous to promote the culture of salmon waters, and to restore them to their former state of fertility.

This lesson shall be confined to two modes of breeding salmon artificially. The first is copied, not *in extenso*, from *Experimental Observations on the Development and Growth of Salmon-Fry*, &c. By John Shaw, of Drumlanrig.—(Edinburgh: Adam and C. Black. 1840.)—Mr. Shaw, as far as my knowledge extends, the first British breeder of salmon artificially, writes thus:—"On the 10th of January, 1836, I observed a female salmon of considerable size (about 16lb.), and two males (of at least 25lb.), engaged in depositing their spawn. The spot, which they had selected for that purpose, was a little apart from some other salmon which were engaged in the same process, and rather nearer the side, although still in pretty deep water. The two males kept up an incessant conflict during the whole of the day, for possession of the female, and were repeatedly on the surface displaying their dorsal fins, and lashing the water with their tails. Being satisfied that these were real salmon, there being at least ten brace of that fish engaged in the same process in the stream at the time, I took the opportunity of securing as much of the ova as I could possibly obtain. This I did three days after it was deposited, the males and female still occasionally frequenting the bed. The method by which I obtained the eggs was by using a canvas bag, stitched on a slight frame formed of small rod-iron, in fashion of a large, square landing-net, one person holding this bag a few inches farther down the stream than where the ova was deposited, and another with a spade digging up the gravel, the current carrying the eggs into the bag, while the greater portion of the gravel was left behind. Having thus obtained a sufficient quantity of the ova for my purpose, I placed them in gravel under a stream of water where I could have a convenient opportunity of watching their progress. The stream was pure spring water. On the 26th February, that is, forty-eight days after being deposited, I found, on close inspection, that they had some appearance of animation, from a very minute streak of blood which appeared to traverse, for a short distance, the interior of the egg, originating near two small dark spots not larger, at that time, than the point of a pin. These two dark spots, however, ultimately turned out to be the eyes of the embryo fish, which was distinctly seen resting against the interior surface of the egg a few days previous to its exclusion. On the 8th of April, which makes ninety days imbedded in the gravel, I found, on examination, that they were excluded from the egg, which was not the case a day or two previous. The temperature of the water at the time was 43 degrees, the temperature of the water in the river 45, and the temperature of the atmosphere 39 degrees. On its first exclusion, the little fish has a very singular appearance. The head is large in proportion to the body, which is exceedingly small, and measures about *five-eighths of an inch* in length, of a pale blue or peach-blossom color. But the most singular part of the fish is the conical bag-like appendage which adheres by its base to the abdomen. This bag is about two-eighths of an inch in length, of a beautiful transparent red, very much resembling a light red currant, and, in consequence of its color, may be seen at the bottom of the water when the fish itself can with difficulty be perceived. The body, also, presents another singular appearance, namely, a fin or fringe, resembling that of the tail of the tadpole, which runs from the dorsal and anal fins to the termination of the tail, and is slightly indented. This little fish does not leave the gravel immediately after its exclusion from the egg, but remains for several weeks beneath it with the bag attached, and containing a supply of nourishment, on the same principle, no doubt, as the umbilical vessel is known to nourish other embryo animals. By the end of fifty days, or the 30th of May, the bag contracted and disappeared. The fin or tadpole-like fringe also disappeared by dividing itself into the dorsal, adipose, and anal fins, all of which then became perfectly developed. The little transverse bars, which for a period of two years (as I have already shown) characterize it as the parr, also made their appearance. Thus, from the 10th of January till the end of May, a period of upwards of 140 days, was required to perfect this little fish, which even then measured little more than one inch in length, and corresponded in all respects with those on which I had formerly experimented, as well as with such as existed at that same time in great numbers in the natural streams."

Mr. Shaw afterwards made additional experiments in ponds artificially constructed, of which the following is a description :—

"The ponds, which are three in number, are two feet deep, and thickly embedded with gravel, while they are at the same time supplied with a small stream of spring water in which the larvæ of insects abound. Pond No. 1 is 25 feet in length by 18 in breadth, and is fed by the stream which debouches into it at the fall F. Pond No. 2 is 22 feet in length by 18 in breadth, and is fed from pond No. 1, at G, where the communication is carefully grated with wire. Pond No. 3 is 50 feet in length by 30 in breadth, and is fed by the stream at F, having no communication with either of the other ponds. The waste water from pond No. 1 is conducted into pond No. 2 through a square wooden pipe, covered at the mouth with a wire grating, the bars of which are about one-eighth of an inch apart. The waste water from pond No. 2 is conveyed under ground, to the distance of 20 feet, in a square wooden pipe, grated in the same manner as the former. The waste water from pond No. 3 passes down a square wooden pipe, two feet deep, covered at the top with wire-gauze, and is conveyed underground in a small covered drain to the distance of 20 feet from the pond. The water of the whole is then left to find its way to the river. To prevent any communication arising from an accidental overflow of the ponds themselves, I raised embankments upon the intersecting walks of two feet in height, so that the several families of fish which the ponds contain can have no access, direct or indirect, to each other. Where the rivulet is divided for the purpose of supplying the several ponds, I have formed an artificial fall in each stream, of a construction to prevent the fish from ascending one stream and descending another. Finally, where the water discharges itself from the ponds, the channels are so secured by wire-grating that it is as impossible for the young fish to escape as for any other fish to have access to them. The whole occupies an area of nearly 80 feet square. My experimental basins being thus prepared, my next object was to secure the fish, the progeny of which were to form the subject of experiment. With the view, therefore, of securing two salmon, male and female, while in the very act of continuing their kind, I provided myself with an iron hoop five feet in diameter, on which I fixed a net of a pretty large mesh, so constructed as to form a bag of nine feet in length by five feet in width. I then attached the hoop and net to the end of a pole nine feet long, thus forming a landing net on a large scale. The weight of the net with its iron hoop being upwards of 7lbs. it instantly sunk to the bottom on being thrown into the water. Being thus prepared with all the means of carrying my experiment into practice, I proceeded to the river Nith on the 4th January, 1837, and readily discovered a pair of adult salmon engaged in depositing their spawn. They were in a situation easily accessible, the water being of such a depth as to admit of any net being employed with certain success. Before proceeding to take the fish, I formed a small trench in the shingle by the edge of the stream, through which I directed a small stream of water from the river, two inches deep. At the end of this trench I placed an earthenware basin of considerable size, for the purpose of ultimately receiving the ova. I then, at one and the same instant, enclosed both the fish in the hoop, allowing them to find their way into the bag of the net by the aid of the stream. In capturing these fish, I considered myself fortunate in securing them by one cast of the net, for, in conducting the experiment of artificial impregnation, it appeared to me to be very desirable that the male should be taken with the female of his own selection at the very moment when they were mutually engaged in the continuance of their species. To take a female from one part of the stream and a male from another might not have given the same chance of a successful issue to the experiment. Having drawn the fish ashore, I placed the female, while still alive, in the trench, and pressed from her body a quantity of ova. I then placed the male in the same situation, pressing from his body a quantity of milt, which, passing down the stream, thoroughly impregnated the ova. I then transferred the spawn to the basin, and deposited it in a stream connected with a pond previously formed for its reception, which, however, I have not considered it necessary to represent in the accompanying plan. The temperature of this stream was 39 degrees ; of the river from which the salmon were taken, 33 degrees ; and of the atmosphere 36 degrees. The skins of the parent salmon are now in my possession. On examining the ova on the 23d of February (50 days after impregnation), I found the embryo fish distinctly visible to the naked eye, and even exhibiting some symptoms of vitality by moving feebly in the egg. The temperature of the stream was at this time 36 degrees, and of the atmosphere 38 degrees. On the 28th of April (114 days after impregnation), I found the young salmon excluded from the egg, which was not the case when I visited them on the previous day. The temperature of the stream was then 44 degrees. The ova, which for some time previous to being hatched, had been almost daily in my hand for inspection, did not appear to suffer at all from being handled. When I had occasion to inspect the ovum, I placed it in the hollow of my hand, covered with a few drops of water, where it frequently remained a considerable time without suffering any apparent injury. The embryo, however, while in this situation, showed an increased degree of activity by repeatedly turning itself in the egg, an action probably produced by the increase of temperature arising from the warmth of the hand."

Mr. Shaw did not breed artificially for the immediate purpose of stocking the river Nith, but to ascertain the growth of salmon-fry *ab ovo*. Having proved that salmon could be bred artificially, he was satisfied with two alleged main results, viz., that salmon-fry, with transverse bar marks, are identical with the "parr," and that they do not attain the smolt, or migratory state, until the age of two years. These results have been disputed by subsequent experimentalists, more particularly by Mr. Young, of Invershin, who maintains that "parr" are adult trout of the smallest variety, and that young salmon (smolts) migrate at the first year, or nearly so, of their age. I am inclined to think that the temperature of the water may hasten or retard the period of migration, but not for so long a period as one year. Mr. Shaw says, "that one or two of each of his three broods assumed the migratory or smolt dress at the age of twelve months." And that he ascribes to the high temperature of water. Mr. Shaw gives an engraving of a two-year old smolt, bred in one of his ponds. It measures 6½ inches in length. I have never seen a naturally river bred smolt longer than five inches, and never one, by any means, so bulky as Mr. Shaw's two-year old smolt.

The last chapter in my *Book of the Salmon* is on "The Breeding of Salmon Artificially." It describes the method carried out by Mr. Young in the bed of the river Shin, successfully, during three consecutive years:—"The first thing to be taken care of in this way of breeding salmon is that the spawning beds, which are to be artificially formed, be supplied, if possible, with water from which the ova are taken. In making experiments on the growth of salmon-fry this precaution is more absolutely necessary than when one is breeding for the sole sake of stocking a river. In all cases it will be advisable, that the spawning and rearing ponds be not fed with water of a temperature widely differing from that from which the spawn has been procured. With these few general remarks, I will transcribe the notes I have received from Mr. A. Young on this interesting and important subject. To give the seed, he says, the same advantages as that naturally spawned in rivers, the artificial breeding-ponds should be erected in the immediate vicinity of, or in the river, and the ponds should be fed by a small stream, or 'lead' taken from the river, so that the temperature and all the conditions of the one may, in every respect, agree with those of the other. At the spot you take the 'lead' off the river, you commence the erection of a wall to shut out the main current. The wall may be built in the river by the side of one of its banks, and its height then is to be greater than the highest flood-marks of the river. In the bottom of the wall, where it takes the 'lead' off the river, an opening or drain-mouth is to be constructed of the width of the current you wish to flow through your ponds inside the defending wall. This opening at the upper end of the wall is to be so framed, that whether the state of the river is low or high, the supply of water to the pond will be neither injuriously diminished nor increased. The drain-mouth, or opening in the wall, is to be secured by a strong iron grating, the bars of which are to be half an inch apart. This grating will prevent the accumulation in the ponds of any thing hurtful to them. The bed of the ponds must be dug up to the depth of about five feet, and they must be nine feet in width, and eighteen in length. Their bottom must be lower by five feet than that of their feeder. The bottom, however, must not be quite flat, but graduated, rising from the end furthest from the head of the current towards the opening or drain-mouth. The necessary inclination can be given to the bottom of the pond by beginning with a layer of gravel one foot thick at the furthest end, and finishing off towards the mouth with a layer of gravel eighteen inches in depth. The bottom of the pond will thus become an inclined plane. The ova are to be deposited at the top of the gradient, where you have finished off with a layer of eighteen inches of gravel, in order that they may have the benefit of sharply running water. The lower part of the inclined plane, or the deepest part of the pond, suits best the fry after incubation. The walls that are to secure the ponds must be strongly built of rough stone. No lime must be used in the construction of the walls, or of any thing connected with the ponds. Every one ought to know the destructive effects of lime upon fish. To secure the ponds from the entrance of the smallest fish, besides the iron grating already mentioned, there must be another fixed inside it of copper wire closely interlaced, so closely as to prevent the possibility of the smallest trout passing through the interstices. If a diminutive trout should enter it would devour the fry as soon as they were hatched. Each end of the pond should be secured in the same way. At the end where the pond water runs out there should be, if possible, a fall into the river, which would effectually prevent the ascension to the ponds of any predatory fish. Some persons have tried artificial breeding in ponds supplied with water from springs and hill-burns, but in such trials no sensible person ought to expect satisfactory results, or, at any rate, results similar to those that would be derived by the use of ponds constructed in salmon rivers, or fed by water directly emanating from them. Both the development of the fish *in ovo* and *ab ovo* depends upon the temperature of the water, and we know that a single frosty night will reduce by many degrees the temperature of rills and rivulets ; whereas the currents of large rivers are little affected by it. Fry hatched in ponds fed by these hill-streams must be stinted in growth—kept in a *status quo* during many weeks—and they can never arrive at the smolt state in the same period of time as fry produced and bred in the waters of rivers. These latter fry are in their natural element—natural in its temperature and in the food, insects, and so forth, it produces. On the contrary, fry bred in ponds fed by springs or hill-burns, are, as it were, subjected to a different climate, strange and unnatural to them, barren, or nearly so, of insects, and foreign to their innate tastes. Their progress in growth, therefore, cannot equal that of fry bred in favorable localities. When the ponds are perfectly formed and constructed, they should be filled with water, and it should be allowed to run freely into and out of them for a few days previously to depositing the spawn in them. This is necessary, in order that the newly-laid gravel may be washed well, the beds properly seasoned, and all mud or alluvial matter got rid of. The artificial spawning-beds must be reduced as nearly as can be to the condition of the naturally formed ones of rivers. The next step to be taken is towards procuring proper spawn for deposition in the ponds. To do so we must watch carefully some natural spawning-ponds of the river at the time when the operations of spawning are going on, and we must capture a pair of salmon that have actually commenced the spawning process. If we do not, we cannot be sure of procuring spawn in a ripe state. We must avoid capturing at random any pair of fish we may see on the spawning-bed, because many consorted males and females are seen hovering about the spawning-grounds several days before they begin depositing their spawn. If from such fish ova are expressed by manipulation, they will be found in an immature state, their pores not as yet open for the reception or absorption of the milt, and expressing it over them will not produce impregnation. On the contrary, when a pair, of course male and female, that have commenced spawning, are captured, their ova and milt will be found in the mature state required, or, at least, a portion of them. A vessel, can, pail, or small tub must be ready, containing a small portion of clean gravel from the river, and as much river water as will cover the gravel, and the seed about to be deposited in it. The female salmon just captured must be held up by the head over the vessel, with one hand, whilst, with the other hand, gentle pressure is made down the belly of the fish. This pressure will cause the expulsion of all the ova that are mature, which will be received in the vessel. The male fish is then to be held and pressed in the same way, which will cause the emission of mature milt into the vessel. The fish are to be returned to their native element, where, if the manipulator be not a rough one, they will speedily recover, and, when the remaining spawn, not artificially forced from them, becomes mature, they will deposit it as if nothing had happened. Having expressed ova and milt into the vessel, it must be shaken so that gravel, water, milt, and ova be properly mixed, and that no ova escape from coming into contact with portions of milt. If any do, they will not be impregnated. On the contrary, the ova that are touched by the milt are impregnated, and, if properly cared for, will, in due time, produce young salmon. I solicit the attention of the owners of rivers to the following great fact :—Salmon-spawn artificially expressed from parent fish, and treated in the manner just now directed, may be conveyed without injury very long distances—from rivers in one country to rivers in another. . . . . To return to our artificial pond, now ready for the reception of the impregnated spawn. It must be imbedded at the head of the pond—at the commencement of the inclination of its bottom, in a small trench about five inches in depth, formed longitudinally with the current, and not across it. The spawn must not be laid all of a heap in the trench, but carefully mixed with gravel all over its bottom, and then covered in with the gravel that has been excavated in forming the breeding furrow. The trench and its covering must be on the slightly inclined plane principle. The gravel with which the trench is covered in must not be pressed down, except very slightly, in order not to prevent the free percolation of the water, which must have full ingress and egress to and from the spot where the seed lies deposited. The action and contact of moving water are essentially necessary to perfect this strange incubating process. Without them ova will be non-productive, for, placed in gravel at the bottom of still, or sluggishly running water, they will putrify, or, to use a generally known expression, they will be 'addled.'"

It will be seen that Mr. Young's method requires no complicated machinery, no spawning utensils or salmon-breeding boxes. He made his beds in the soil of an old mill-race in the river Shin, and took his spawn from that river. Of course the temperature of the waters of the spawning-beds and of the river was the same. His artificially bred salmon fry assumed the silvery, migratory, smolt coat at the end of twelve months. This proves only that the Shin salmon fry become smolts at a year old, but it does not prove that to be the case in *all* rivers. I think it is so in the great majority of salmon rivers. I should very much like to see impregnated salmon ova from the Shin placed in the river Nith, and *vice versa*. If done, it would go far towards enabling us to come to a conclusion as to the age of smolts.

EPHEMERA.

Jan. 27.

LESSON V.

On the 8th December last I commenced these lessons; I now conclude them. They have given rise to most important communications to this journal, on the subject of breeding salmon artificially, from a Lancashire gentleman, who signs himself "Salmo," and from a Scotch gentleman who has for years written on all that relates to salmon, under the signature of "Y." They have led to other writings on the subject in the provincial papers, by far the most important of which are those contributed to the *Kelso Mail*, now printed in a pamphlet form, by Mr. Thomas Todd Stoddart, the author of that excellent book, *The Angler's Companion to the Rivers and Lochs of Scotland.* At the end of this lesson I shall make extracts from Mr. Stoddart's pamphlet. In the course of my lessons I have shown that salmon breed in the shallows of rivers, in nests they form in the gravel—that the female deposits her ova in them, and that immediately afterwards the male impregnates the ova by shedding his milt upon them—that the ova are hatched on an average in 120 days—that the fœtus so produced does not assume the perfect fish form until one month old—that it is then a salmon fry, and so continues to its twelfth month, when it becomes a smolt, with silvery coat, its length five inches or so, and its weight two or three ounces—that it then migrates to salt water, feeds therein, and in the course of three or four months becomes a grilse or maiden salmon, weighing not unfrequently 8lb. (rapidity of growth most marvellous and almost incredible !)—that in this grilse state it returns to its native river, spawns in due season upon its fords, and not long afterwards migrates for the second time to sea—that having fed there for two or three months, it becomes an adult salmon, weighing 12, 14, or 16lbs.—that it again returns to its native river, deposits in the spawning beds some 10,000 or 15,000 ova, which are impregnated by the male, and give life, in all probability, to 7,000 or 8,000 salmon—that salmon will migrate and immigrate every year, until they are captured, increasing in size, not in the same ratio as in their first years, annually, and breeding annually. I have shown how salmon are to be bred artificially, and I have noted many of their minor and curious habits for the satisfaction of the young naturalist and angler. Nothing remained for me to do but to show how salmon might be best preserved. I was just about to do so, when I received the communication, printed underneath this, from "Salmo" on the very same subject. He has forestalled my suggestions, and so ably shown how salmon are to be increased and preserved, that I deem it for the present totally unnecessary to write a single word more upon the subject. I am not, however, going to close my lessons abruptly. I have something very singular to communicate to my readers—something that will induce them to fancy that the habits of salmon are not the same in all rivers and countries. I have occasionally fancied so myself, and supposition is now almost converted into belief. By what means ? I will show you, reader, in a few minutes.

There have appeared lately two large imperial octavo volumes, beautifully printed, and most profusely, ornamentally, and usefully illustrated, published by Mr. Bentley, New Burlington-street, publisher in ordinary to her Majesty. The title of this superb, interesting, and instructive work is, *Scandinavian Adventures during a Residence of Twenty Years*, by L. Lloyd, author of *Field Sports of the North*. The title-page furthers tells, and truly tells, that the volumes "represent sporting incidents, and subjects of natural history, and devices for entrapping wild animals ; with some account of the northern fauna." A large portion of the first volume is occupied with the history, habits, and modes of capture of the fish of Scandinavia. For the present I shall confine myself to a portion of what the author writes about salmon. Chapter VI. is devoted to that fish, and commences thus :—"The natural history of the salmon tribe having of late years excited much interest in England, I cannot do better than to devote a chapter to some remarks, the result of an attentive study of their habits for several consecutive years, recorded by my gifted friend and countryman, Mr. Alexander Keiller ; observations which, I doubt not, will be interesting even to the unscientific and general reader." The observations were made on the river Save, a tributary of the Gotha, near Jonserud, where the water is invariably clear. They were made from a moveable observatory, lantern-shaped, suspended by means of ropes and pullies from a pole fixed in the bank and inclining over the river. This observatory is large enough to hold a man, and a drawing of it represents Mr. Keiller seated in it, and observing through its bottom the fish that swim or perform other operations beneath him. You have seen a bale of goods suspended from a crane over the hold of a vessel. Imagine the bale a large glazed frame, with a bottom strong enough to sustain the weight of a man, the hold of the ship the bed of a river, and Mr. Keiller sitting within the glazed frame, and then you will have a notion of his ingenious observatory.

Let us now see what he saw from his watch-box hanging over and a little above the river. "Salmon," he says, "are pretty abundant in the Save. The fishing produced, including grilse, about three thousand pounds weight annually. Many fish were taken in weirs, others in nets, or by the rod. The larger salmon always appear first in the spring ; as the summer advances, the fish are much smaller ; but in the autumn heavy fish again show themselves. These are not *fresh run*, however ; at least they are somewhat discolored, from which it is to be inferred they have been lying either in brackish water or in the deep pools below." In the British rivers, generally speaking, the same thing occurs. The small summer salmon are with us, for the most part, grilse. What will Mr. A. Young, "Salmo," Mr. Stoddart, my Worcester friends, Messrs. Allies, George, and Flinn, and the scientific Gottlieb Boccius say to the following mode of breeding? I am all agape as I copy it:—"Salmon commence spawning in the Save the first days in November, and continue through the month. The female deposits her eggs in comparatively still, shoal water, from six to eighteen inches in depth, immediately above a rapid. She selects such a situation for the following reasons: Comparatively still water in preference to a current, because otherwise the exertion of retaining her position, and spawning combined, would be too much for her powers; a shallow, instead of a pool, that she may be secure from the sea-trout and other fish, which, if in deep water, would congregate about her to prey upon her eggs; and lastly, that her ova, in dispersion, may be carried by the gentle stream to a secure resting-place amongst the stones below." In this country the choice of shallow water, by breeding salmon, is attributed to a far different instinct, viz., that the ova, or rather the water in which they are, may be subjected to the influence of atmosphere and light, and be more oxygenated than they would be if deposited in still water. A current is considered absolutely necessary for their development. Now comes a statement totally opposed to our salmon beds—"hills" and "ridds," as we call them :—"It is commonly supposed that, in conjunction with the male, the female salmon scrapes a hole or furrow in the bed of the river [they certainly do so in the rivers of the United Kingdoms, England, Scotland, and Ireland; I do not know how they manage connubial matters in the Principality], in which to deposit her eggs, and that afterwards, and as a protection from their numerous enemies, they cover them over with gravel; but such is not the fact in the Save. The male has nothing to do with this part of the work; and the ova, instead of being dropped into a cavity, are deposited on a comparatively smooth surface. Whilst in the act of spawning, the female retains her natural position. Her belly is near the ground; at times, indeed, probably to rest herself, actually touching it. The process of dropping her eggs appears to be slow. When a few are collected, she turns on her side, waves the flat of her tail gently towards the roe, but lifts it up again with great force, by which such a vacuum is caused as not only to raise the eggs from the ground, and thus to distribute them in the stream, but to throw up a mass of dirt and stones, the latter not unfrequently of very considerable weight. As the mere distribution of the ova would require only a slight wave of the tail, it appears that the violent lunge is for the express purpose of disturbing and muddying the water, thereby to conceal the eggs, in a degree at least, from their numerous enemies lying in wait below. When spawning has once commenced, it seems that the male can no longer retain his milt, nor the female her roe, the emission continuing under all circumstances. This has often been noticed, even long after death. The female salmon leaves the spawning-bed many times during the day, and makes little excursions about the river, generally into the dead water above. At times these trips are somewhat extended—say to a distance of some seventy or eighty paces. "But," said Mr. Keiller, "as from my elevated position I could watch all her movements, I feel perfectly confident that, during her absence from the spawning-bed, she never in any way comes in contact with the male fish. I am at a loss to understand the cause of these trips. At times, I have thought it is for the purpose of resting herself after the fatigue or exhaustion of spawning; at others, I have imagined it to be a special provision of nature; for if her original position were a bad one, and she were to remain stationary, all her roe would be destroyed; whereas, by occasionally moving, as she does, about the stream, and dropping her eggs as she goes, some of them, at least, are pretty certain to find shelter." The specific gravity of the roe is but little greater than water; when once, therefore, in motion, unless intercepted, it will float a considerable distance down the stream. A large portion of the eggs are of course devoured, but the remainder find their way into crannies, and under stones inaccessible to an enemy. From the slow manner in which the salmon spawns, it might be thought on the first view of the subject that a large portion of the eggs in the body of the fish were in an *immature* state; but such is not the fact. [It is the fact in this country, and therefore a female salmon of large size is obliged to continue the operation of spawning for several days. The ova next the throat are *immature*, and cannot be expelled except by violent pressure. Repeated experiments have proved that they will not absorb the milt, and therefore cannot be impregnated. Mr. Keiller's experiment, proving the contrary, is very extraordinary. I am bound to believe it, but cannot account for it. Can you, "Salmo" and "Y?"] To prove this, Mr. Keiller once took the roe in a mass from the belly of a salmon recently captured, divided it transversely into three equal parts, and applied to each the needful quantity of milt. [What, without separating the ova?] In due time the several portions produced fry, though it is true that the portion taken from the upper part of the belly when the eggs were of a somewhat less size, was less productive than the other two. [To be sure it was, because the unproductive ova were *immature*.] . . . . At the tail of a spawning ground the work of a single salmon—or at all events never occupied by more than one at a time —there is, towards the close of the season, an immense accumulation of gravel, stones, &c.—occasionally, indeed, a good English cartload. [What causes the accumulation of these stones? Surely there must have been some digging, or excavating process, done by the salmon to form such a cairn.] What with ice and floods, however, not only is this heap in great part carried away, but the very cavity [Who, or what formed the cavity ?] from whence it came, often of great extent, is so filled up, that by the succeeding summer the bed of the river has assumed nearly its usual appearance. . . . . It has been shown that, whilst the female is spawning, the male is stationed some ten feet in her rear. Again, at a respectful distance behind him—say twelve or fifteen feet, but still in a direct line with the female—a lot of trout, sea-trout, and other fish are always posted, in readiness to pounce on the eggs, when the female starts them adrift with her tail. On the appearance of the several clouds of dirt, it is amusing to see them all scurrying into the thick of it, and following the ova down the stream. It has never been observed that the female has a liking for one male more than another; but it has been repeatedly noticed that some one male in particular occupies the same spot. At some little distance to the right and left of this male, two or three other males are usually to be seen, and much of his time is occupied in keeping these interlopers at a distance. His charges against them are most vigorous and determined, and so frequent that he is seldom stationary for a minute together. This almost incessant motion of the male seems a special provision of nature ; for, were he to remain still, only that portion of the ova which passes over him would be impregnated, whereas, by moving so much about, his milt becomes distributed, in a manner, over the whole stream."

I must say that, if this is the breeding process pursued by Swedish salmon, it is a very improvident one. I cannot see how a tithe of the ova can be impregnated by it, since so many must escape coming in contact with the milt, so many must be washed uselessly away, and so many eaten by trout and other predatory fish. I can vouch for it that our Sutherlandshire salmon are much more "cannie" in their manner of accouchement, &c. An illustration, very curious, is given of three pair of salmon in the act of spawning. We see one female salmon on her belly, about to deposit her ova, a male lower down in a line with her, his head turned aside, "casting a jealous glance at an interloper." Mr. Lloyd, carrying on the description of the diagram, says :—"Second pair in the centre—female on her side, in the act of distributing her ova (when shed, with her tail), the male passive, and the fry (predatory trout, &c.) revelling in the passing cloud. Third pair to the right —the female passive, the male seizing a poacher on his manor, in which interval, it will be observed, an intruder takes advantage of the liege lord's absence, and is about assuming his place. The zigzag lines represent the manner in which the milt of the male salmon, according to his peregrinations, become distributed over the whole river." A glance at the illustration, of which the above is a written description, will at once show the extraordinary and extravagant way in which Mr. Keiller's salmon spawn. It may be recollected, that in my writings I more than once stated that I could not account for the cartilaginous, upright excrescence found in the under-jaw of the male salmon during the spawning period. The following is a singular opinion as to the use made of it :—" It is the commonly-received notion, that the hook on the lower jaw of the male salmon is for the purpose of enabling him to assist the female in forming a hole in the bed of the river, for the deposit of her roe. But such Mr. Keiller convinced himself is not the object for which it is designed. In his opinion, it is intended to prevent the males, which in the spawning season are most pugnacious, from killing each other, for when the jaws of even a 25-pound fish are distended to the utmost, the hook is so much in the way, that the opening in the front of the mouth will admit little more than the breadth of a finger, and, consequently, he cannot grasp the body of an antagonist. Indeed, were he able to do so he would soon destroy him." Swedish salmon appear to have all the fighting propensities of that most disinterested of royal soldiers, Charles XII. ; *exempli gratiâ*—" In the breeding seasons the contests between the males are incessant and desperate. Mr. Keiller repeatedly noticed an immense salmon charge another with such thorough good will as to throw him fairly out of the water. As it is, their battles are bloody enough ; not only are the fish observed to be gashed in every direction—probably by their side teeth, for those in front, or on the tongue cannot be brought properly into play, owing to the hook—but with large pieces of flesh and skin actually hanging down their sides. At the close of the season all the males are covered with scars. Unless one has seen the fish at this time, it is difficult to conceive his mutilated condition, and it appears certain that were it not for the hook not more than a single male salmon would leave the spawning ground alive. But it is the males alone who, at the termination of the spawning season, are thus seamed with scars. Another evidence, were such wanting, that the injuries have arisen from combats between themselves, for, were the wounds inflicted by otters, as many imagine, the females would be equal sufferers with the males, which is not the case." All male animals—man among the rest—fight and make fools of themselves to be the lords and masters of females. The jackass is a ferocious ass in this respect, and the stag or hart will knock to smithereens his antlers, and dash his Turkish brains out in a fiery contest for a harem of hinds. Yet, according to the bulletin of Mr. Keiller, the salmon is the most sanguinary sultan of them all. I have done for the present with quotations from *Scandinavian Adventures*, the best work by far ever written on the field sports of Northern Europe.

I have, in another column, noticed it more prominently. The habits of the salmon, as above described, are after Mr. Alexander Keiller. Mr. Lloyd does not always coincide with the conclusions of his friend. Mr. Keiller is of opinion that salmon fry pass two summers in fresh water previously to their first migration to the salt-water feeding grounds.

I shall now carrry out my promise, and notice Mr. Stoddart's pamphlet on *The Artificial Breeding of Salmon in its connection with the Tay and the Tweed.* He does not think artificial breeding necessary in the Tay, and of the mode in which it is now being tried at Stormontfield, on that river, near Perth, he has by no means a favorable opinion. He makes the following calculations to show that there are abundance of spawners in the Tay and its tributaries, and that the assistance of 400,000 impregnated ova, placed in the artificial breeding boxes at Stormontfield is not wanted, and is paltry in comparison with the 150,000,000 ova naturally deposited annually in the Tay and its tributaries :—

"I think I am in no degree overshooting the mark, when I set down the annual number of salmon and grilse which spawn in the Tay and its tributaries, at 30,000. The fact that the average number of salmon, independent of grilse altogether, which is captured annually in Tay, approaches the figure fixed upon, is of itself good security for the spawning to that extent, in any previous year. In 1846, for instance, there were taken by the net on Tay and

Earn, below the mouth of the Islay, upwards of 34,000 salmon and 30,000 grilses. The 34,000 salmon, according to certain well-established theories, must consequently have spawned in Tay and its tributaries, some as grilse, some as salmon of inferior growth, during the fence season of 1845. A supposition, therefore (if it deserves that doubtful name), which confines the number of spawning fish on these rivers to an average of 30,000, cannot be held as an outrage upon truth—nay, is greatly within limits. These 80,000 fish, we shall suppose, according to the notions set afloat by Mr. Young and others, are equally matched in respect to sex; that is to say, that out of the whole number, 15,000 are spawners, and the other 15,000, milters or he-fish. To each spawner I assign an average weight of 10lb. A ten-pound breeder, it has been ascertained, yields about 10,000 ova, certainly not fewer; consequently the number of ova cast, in the spawning season, on the various breeding grounds referred to, may be computed at 150,000,000. How much of this large amount of spawn are we entitled to suppose is brought to life, becomes distributed over the rearing grounds, attains the smolt size, and, assuming its plumage, eventually finds its way into the sea or firth?"

Mr. Stoddart computes the result of the above number of ova at 20,000,000 smolts, having allowed for every sort of destructive casualty. He calculates that the 400,000 ova deposited in the breeding boxes will produce 300,000 fry—a very high calculation indeed—and he then says:

"The artificial nurseries or rearing grounds in process of construction are, I fear, much too limited in size to afford accommodation to the number of fry anticipated from the boxes. I shall assume that they occupy an area of five acres, although I have reason for believing that not one fourth of that extent of surface is intended to be put into requisition. Five acres of ground converted into ponds or reservoirs for the reception of 300,000 fry! Five acres, in a state of artificiality, subjected, possibly enough, to the influence of a stream or diversion of water passing through their several divisions, and so preventing them from becoming stagnant, but totally devoid of the advantages possessed by the natural rearing grounds of Tay, in the shape of shelter and sustenance! Are we justified in taking it for granted that so limited an extent of nursery will suffice for the wants, assist the growth, and do justice to the condition of 300,000 young salmon, during the space of an entire twelvemonth?—in other words, that a single square yard of artificial rearing ground, be the depth of water what it may, affords ample enough accommodation for twelve smolts and the fractional part of a thirteenth, throughout the whole period? I certainly think not."

It seems that the fry artificially produced near Perth are to be artificially fed for one whole year.

Upon this senseless project Mr. Stoddart sensibly remarks :—

"As to the hand-feeding proposed, it may be all very well to apply it to the bringing up of such fish as carp and tench, but to carry it into successful operation with young salmon, will, I fear, turn out a work of impracticability. An expensive one, at any rate, it is acknowledged to be—the most costly item in the whole series of experiments. Why, therefore, pursue it at all ? Why not rest content with the hatching process, and allow the infant fry, after its full development, convenient access to the river ? A sluice or run in communication with the boxes towards the Tay will effect this object. In that case —were the experiment allowed to rest at this juncture, the object of it may possibly to some extent become attained. It never can, by prosecuting it farther in the way proposed—that is by cooping up the fry in mere tanks or ponds of very limited dimensions, and by hand-feeding them with such unnatural aliment as chopped liver."

If the fry hatched at Stormontfield are confined in nurseries for a year, or until the migratory period, the experiment will be a failure. When a month old they should be turned into the river, and they will take care of themselves, migrate naturally, and all that escape destruction return as grilse. I am of opinion that confined and hand-fed fry, if confined and fed for the space of a year, will not return, if they do migrate to sea, from it.

Mr. Stoddart finds the stock of fry very scanty in his own river, the Tweed, and he proposes something like artificial breeding in it to increase the stock, which appears as necessary to be done in the border rivers as it is unnecessary in the abundantly stocked Tay. He writes :—

"The opening of Tweed on the 15th of February, as is well known, has of late years been signalized or rather rendered notorious, by an immense slaughter at many of the netting stations not only of kelts, but of baggits and kippers to boot. Last year, as I have been given to understand, scores of ripe spawners were captured during the opening week below Tweedmill and in the vicinity of Twizel. I have known to the amount of 80 she-fish, all large and primed with ova, having been taken in a single day, from the Tweed, on a similar occasion, and there is every reason to believe that the termination of the present fence time will be followed up, as usual, by extensive massacres of unspawned salmon and grilses. Now, what I propose is this, that the proprietors or parties holding salmon fishings on Tweed should instruct competent persons to attend the various netting stations at the opening of the season, for the purpose of expressing, collecting, and inoculating, when opportunity offers, this great annual wastage of spawn—for the purpose also (not of stowing it away in wooden boxes, over which an artificial run of water shall be directed), but of committing it to 'ridds' formed with the shovel, hoe, or plough in the bed of the river itself, there to await, as a matter of common certainty, its being brought to life. . . . . Immediately on the expiry of close-time, the nets and cobles are set in motion. A few shots determine, in most cases, the contents of the river, near the station where these are made. For every clean salmon taken during the first fortnight, in Tweed, there are at least a dozen kelts and four or five unspawned fish generally in a very forward or mature state. These are secured, as a matter of course, during the ordinary endeavors made by the fishermen to bring the net into contact with something better. There is no cost or extra labor therefore required, in order to obtain the spawn. The attendance of one or two of the ordinary river police at each likely station, during the first three weeks of the season, is all that is needed in order to collect the ova and conduct the inoculating process. This, under the instructions of Mr. Mitchell, the active superintendent of that body, any one accustomed to the handling of salmon can accomplish. Say the ridds, by permission, are formed with shovel or plough at the Monk's Ford, betwixt Old Melrose and Dryburgh—nothing more is necessary than to forward the spawn to the Newtown station, and thence conveying it to the ford in question, mix it up with gravel or river sand, and commit it to the ridd; all which may be done within two or three hours of its being taken from the fish. Can any thing be simpler or less expensive? Well, mark the results. Here is a quantity of ova which never would have a chance of being brought to life (for even supposing the baggits themselve are returned alive to the river, the disturbance occasioned by the constant plying of the nets on or near the spawning grounds, and the liability such fish incur of being retaken, over and over again, make miscarriage almost inevitable)—these ova, down to a single pellet, are rescued from certain destruction and buried with extreme care in a choice portion of the river, where no ordinary calamity can possibly overtake them. The result will be, that almost all the ova so inhumed will come to life ; and say that they form the supply from only two hundred baggits, each baggit yielding a trifle beyond 10,000, we have at once, added to the natural resources of the Tweed, a hatch or brood consisting of 2,000,000 of fry, all vitalized at the expense of a few pounds sterling. Of these fry, nearly one-half, without any additional cost whatsoever, is likely to attain the smolt stage, and allowing that only a single individual out of two hundred find its way back to Tweed, in the shape of a grilse, the annual produce of the river undergoes an increase of nearly 5,000 available fish. Let Mr. Ramsbottom's system of breeding and rearing, at one farthing per smolt, match this if it can."

I am strongly of opinion that the Tweed Commissioners would act wisely by paying prompt, substantial attention to the feasible suggestions of Mr. Stoddart. It is a great pity to allow poor old Tweed to go to rack and ruin. We London salmon anglers shall lament over her irremediable decay, for, apart the beauty of her scenery, she may, if Mr. Stoddart be listened to, be rendered the best nearest salmon river to the English and Scottish capitals.

For this year, at least, I have done with the natural history of salmon. Let not my drying my pen on the subject prevent "Salmo," "Y.," and others from keeping theirs in working trim. For March 19th I shall write, with this motto from the *Captives* of Plautus,

"Nunc hoc animum advortito, ut ea quæ, sentio pariter scias,"

my first angling lesson for the season.

EPHEMERA.

Feb. 24.

ON THE BREEDING AND PRESERVATION OF SALMON.

[The following article on the Breeding and Preservation of Salmon, is the communication signed SALMO, published in *Bell's Weekly Messenger* of March 5, 1854, and referred to in the Fifth Lesson of EPHEMERA, on Salmon Breeding, published in the same number of that journal.]

MR. EDITOR: We all remember the sensation produced by a late and now lamented secretary for Ireland, when, remonstrating with the landowners of that suffering country, he reminded them that "*Property has its duties as well us its rights.*" Every man is conscious of the truth of this remark, which has now passed into a proverb, but few men could have said it so pithily and so well. I intend to borrow this maxim as the basis of the suggestions I have to offer as to the best means of preserving and increasing the breed of salmon, of which it is impossible to magnify the importance. As an article of food, it is beyond comparison the most valuable of fresh-water fish, both on account of the delicacy of its flavor, and the numbers in which it can be supplied. By wise and prudent legislation it may also be rendered cheap, and accessible to the family of an ordinary artisan.

The salmon requires only two conditions to enable her to mutiply and increase indefinitely, viz., an undisturbed breeding-ground in fresh water, and an uninterrupted feeding-ground in the sea. How happens it, then, that these two simple wants cannot be conceded? I will tell you in a few words. The salmon, being a migratory fish, is continually changing its habitation. To-day it may be in the sea—to-morrow in the river—and in a few days afterwards at the top of some remote mountain stream. John Graball, who owns the lower part of the river, reasons thus with himself:—"All the salmon which comes into the river, must pass through my liberties before they can be taken by any one else, and as salmon is now selling at 3s. a pound, I will let none of them escape if I can prevent it. Now's my time to reap the harvest; what I leave, any one else may catch if he can." The same motives and the same feelings actuate John's neighbors higher up the stream, until you reach a point at which the fish arrive so late in the season, or in such small numbers as not to be worth the expense of men and nets. The proprietors of the upper rivers then, with the exception of a few occasional fish taken by the rod, derive no benefit from the salmon. They have the pleasure of preserving them, finding them breeding ground, watching over them during the most critical parts of the year, and, when these duties have been fully and conscientiously discharged, they have no other reward than the satisfaction that they have acted as good citizens, and have merited the approbation of the public. However, it sometimes happens that one of these proprietors is not over scrupulous. He will take fish out of season, either with rod or net. He will spear the spawning salmon for the sake of the roe, and make a profit by selling it to the retailers of fishing tackle, who dispose of this tempting lure in large quantities to ambitious anglers. Perhaps John Graball gets to hear of this illicit traffic, and pays a visit to the unlicensed poacher, and threatens with information and fine, or possibly with imprisonment, if the money be not forthcoming. Our friend, however, does not let Mr. Graball have all the talk to himself. We may suppose him making some such answer as this :—"I'll tell you what, Mr. Graball, if we come to balance accounts fairly between you and me in the matter of these fish, you have the best of the bargain by long odds. You only own about half a mile of the river, and yet you contrive to get a profit of £500 or £1,000 a year out of it. You never breed a fish. All yours is deep water, without a yard of breeding ground in it. You never feed a fish, for they all come out of the sea ready fed, and ready fattened to your hand. Now, look at me. I have five miles of water running through my estate, and brooks and rivulets without number. From November to January my waters are crowded with fish, salmon, and grilse, and sea trouts; but they are worth nothing to me. They come here to breed, and I never see a fish fit for the table. See the thousands, aye, millions of young fry that leave my waters for the sea every year, and yet I have no benefit from them. You catch them all at the mouth of the river, and make a fortune by them, whilst they do not return me the value of a shilling. I breed them; her Majesty, as Queen of the Ocean, feeds them, and you catch and sell them, and yet you grudge me a trifle of salmon roe, which I take more for the name of the thing than for any great profit to be made by it. But I tell you what, John Graball, now that you have broached the subject, I intend to be plain with you, I will not consent any longer to breed salmon for you, unless I am to have a share of the profits. I will throw the river open to everybody who chooses to come, either with rods or nets. I will watch the fords at night, and I will spear every salmon I can find. I will do at the upper part of the river what you do at the lower part. I will take every fish I can lay my hands on, and make all the profit I can by it. Yet I am not an unreasonable man, John; I admit that property has its duties, and it is my duty to preserve the salmon for the public good. But there is another side of that question, and property has its *rights* as well as duties; and if I do my duty by preserving the salmon, I claim an interest in the salmon as my right. It is of no use to tell me that the law will prevent or punish me for taking foul fish, or destroying spawn. When the law is palpably unjust and one-sided nobody cares for it, nobody respects it. The law, to be respected and cared for, should deal equally with the rights of property, and the duties of property; and if it enforce the one and neglect the other, it will only generate contempt and apathy. Now I'll tell you what I'll do, John; I'll join with you in calling a meeting of proprietors from the top of the river to the bottom, and we will try to make a bargain that shall be for the benefit of all. Let us form ourselves into a company, and preserve the waters carefully at our joint expense, and if there be any profit in the fisheries, let us join in the profit. You cannot have the salmon if I refuse to breed them, and I do positively refuse any longer to keep a brood farm, and let you have all the young stock. If you agree to this offer, we can breed ten times as many fish as we do at present, and we shall all do our best to preserve them. As matters now stand, I care nothing about them; I have no interest in them beyond the pride of having salmon in my waters; but you know, John, that pride will not pay rates and taxes. Well! is it a bargain? Are we to be partners or antagonists? choose for yourself."

You have here, Mr. Editor, if not the words, at any rate the sentiments, of all those river proprietors who are debarred, by their distance from the sea, from ever taking a salmon in a seasonable condition. There are many of them who incur considerable expense in watching and preserving their waters, and yet derive no pecuniary advantage from their care and expenditure. This is manifestly unjust, and this injustice is the main cause of the diminution in the numbers of salmon. Unless the breeding-grounds are superintended with great care and vigilance, the only source of supply is cut off. As the salmon diminishes in numbers, the owners and lessees of fisheries are compelled to use greater diligence to capture the few that present themselves, and these two causes continue to operate in a gradually increasing ratio, until the breed must and will become extinct, unless some prompt and efficient remedy be adopted.

What, then, is the remedy? Simply this: give every proprietor of land on the banks of a salmon river, a legal right to a participation in the profits of the fisheries. Offer him every inducement, by that most influential and seductive of all motives, self-interest, to become a breeder and preserver of salmon. By the same agency, you deprive him of every motive to kill unseasonable fish, or destroy their ova. He has a claim to this participation by natural and undoubted right, and, until this concession is made to him, all other projects must and will fail. *Property has its rights as well as its duties,* and it is one of the rights of property, that he who breeds the fish, and feeds the young, until they are strong enough to migrate to the ocean, ought to share in any profit to be derived from the capture of these fish, on their return to the river. To deny this proposition, would be a libel upon the common sense and the primitive instincts of man, which never fail to appreciate, though they often fail to analyze, a glaring and palpable injustice.

You may ask me how I propose to accomplish this object. In the first place, I would put an end to all conflicting rights on a salmon river by making the proprietors a corporate body, and placing the whole under one management. Take, as an example, the river Ribble. I would constitute the proprietors a company, under the name of "The Ribble Fisheries Company." Every proprietor should be a member of this body, and entitled to a voice in its management, as in a railway or other company constituted by an act of Parliament. By this means we should get rid of many jealousies, where there are conflicting rights, and we should have salmon at the lowest cost, because there would only be one set of nets, one set of men, and one board of management. I would prohibit, under heavy penalties, the use of a net in any part of the river, by any but the constituted authorities, unless on a written application for some experimental purpose, to be submitted to the board, and conducted under the superintendence of a conservator of the river. By this plan every landowner on the banks of the river, or of any tributary stream to which salmon resort for breeding, would become a shareholder in the company, and entitled to participate in the profits. The only practical difficulty would be to determine the proportions in which the profits should be divided between the proprietors of the upper and lower parts of the river. This is a matter of detail which might easily be arranged, as questions of tenfold more difficulty occur every year in the conduct of ordinary corporations. It may be asked, would I make the adoption of this scheme compulsory on every salmon river. Certainly not. I would first pass a general act of Parliament prohibiting the use of stake nets or any other engine or device of any description whatsoever that was fixed and self-acting in any river, or estuary of a river, frequented by salmon. The close days might be left to the discretion of the magistrates assembled in quarter sessions, as provided by statute now in force. It would then be necessary to prohibit the use of any net in a salmon river under a certain sized mesh, so as to prevent the taking of all young fry, or fish, under a given weight. I think these provisions imperatively necessary in any public act. It is impossible to say how many thousands and millions of young salmon are destroyed by these pestilent inventions. Take as many as you can with the rod, but let no man, under a penalty, take with a net, in a salmon river, any fish less than five pounds weight. This would simplify the matter very much, and would injure no one, as the real mercantile value of a fishery is determined almost altogether, by the weight and number of the salmon taken. Let any man consider for himself the amount of mischief done by taking so many thousands of salmon in their first year of immigration, averaging from 1lb to 5lb, instead of giving them another season's grace, to return valuable prizes of 7lb to 14lb each. I speak within compass when I say that this single regulation would quadruple the number of salmon in the second year of its trial. After making these provisions I would give power to a certain number of the proprietors, by mere petition to the House of Commons, to incorporate the fishery, and erect themselves into a company, with power to levy rates, to appoint directors and other officers, in the same manner as any other public body which derives authority from an act of Parliament, the powers and functions of this body and its officers to be defined by the public act. I have here given an outline of a scheme which, I feel satisfied, would meet the urgent and immediate wants of the salmon-loving public. The general good would be consulted by strict enactments against the present vile and destructive practices. Private interests would be respected by recognising a property which at present is ignored by the law, although it is recognized by conscience and common sense ; conflicting claims and mutual jealousies would subside under the influence of a joint and common interest ; poaching and unseasonable fishing would rapidly abate, if not disappear altogether, by uniting in the closest of ties all persons interested in the protection of the fish ; and the salmon would enjoy, for the first time in the memory of man, the full opportunity of developing all her resources, and of proving how much she has been libelled by those feeble drivellers, who have voted her incompetent to discharge the functions which constitute the chief end and object of her existence. Give us these provisions, ye men of tanks and incubators! ye philosophers of conjecture and romance! ye novelists and theorizers! Grant to the salmon one-half the chance you give to your pigs, your cattle, your poultry, or your game, and let us see whether she forms the only exception to that noble and beneficent ordination of Providence which has formed every animal perfect in its kind, and has given it the instincts and the capacity to accomplish worthily and effectively the purposes for which it was created.

Feb. 18th, 1854. SALMO.

THE END.

History of the Romans under the

Empire. By CHARLES MERIVALE, B. D., late Fellow of St. John's College. 7 vols., small 8vo. Handsomely printed on tinted paper

CONTENTS:

Vols. I. and II.—Comprising the History to the Fall of Julius Cæsar.
Vol. III.—To the Establishment of the Monarchy by Augustus.
Vols. IV. and V.—From Augustus to Claudius, B. C. 27 to A. D. 54.
Vol. VI.—From the Reign of Nero, A. D. 54, to the Fall of Jerusalem, A. D. 70.
Vol. VII.—From the Destruction of Jerusalem, A. D. 70, to the Death of M. Aurelius.

This valuable work terminates at the point where the narrative of Gibbon commences.

. . . "When we enter on a more searching criticism of the two writers, it must be admitted that Merivale has as firm a grasp of his subject as Gibbon, and that his work is characterized by a greater freedom from prejudice, and a sounder philosophy.

. . . "This history must always stand as a splendid monument of his learning, his candor, and his vigorous grasp of intellect. Though he is in some respects inferior to Macaulay and Grote, he must still be classed with them, as one of the second great triumvirate of English historians."—*North American Review, April*, 1863.

Practice in the Executive Department

partment of the Government, under the Pension, Bounty, and Prize Laws of the United States, with Forms and Instructions for Collecting Arrears of Pay, Bounty, and Prize Money, and for Obtaining Pensions. By ROBERT SEWELL, Counsellor at Law. 1 vol., 8vo. Sheep.

"I offer this little book with confidence to the profession, as certain to save lawyers, in one case, if they never have any more, more time and trouble than its cost. To the public generally, the book is offered as containing a large amount of useful information on a subject now, unfortunately, brought home to half the families in the land. To the officers and soldiers of the Army it will also be found a useful companion; and it is hoped that by it an amount of information of great value to the soldiers, and to their families at home, will be disseminated, and the prevailing ignorance respecting the subject treated of in a great degree removed."—*Extract from Preface.*

Hints to Riflemen.

By H. W. S. CLEVELAND. 1 vol., 12mo. Illustrated, with numerous Designs of Rifles and Rifle Practice. Cloth.

"I offer these hints as the contribution of an old sportsman, and if I succeed in any degree in exciting an interest in the subject, my end will be accomplished, even if the future investigations of those who are thus attracted should prove any of my opinions to be erroneous."—*Extract from Preface.*

Round the Block.

An American Novel. With Illustrations. 1 vol., 12mo.

"The story is remarkably clever. It presents the most vivid and various pictures of men and manners in the great Metropolis. Unlike most novels that now appear, it has no 'mission,' the author being neither a politician nor a reformer, but a story teller, according to the old pattern, and a capital story he has produced, written in the happiest style, and full of wit and action. He evidently knows his ground, and moves over it with the foot of a master. It is a work that will be read and admired, unless all love for good novels has departed from us; and we know that such is not the case."—*Boston Traveler.*

The History of Civilization in

England. By HENRY THOMAS BUCKLE. 2 vols., 8vo.

"Whoever misses reading this book, will miss reading what is, in various respects, to the best of our judgment and experience, the most remarkable book of the day—one, indeed, that no thoughtful, inquiring mind would miss reading for a good deal. Let the reader be as averse as he may to the writer's philosophy, let him be as devoted to the obstructive as Mr. Buckle is to the progress party, let him be as orthodox in church creed as the other is heterodox, as dogmatic as his author is sceptical—let him, in short, find his prejudices shocked at every turn of the argument, and all his prepossessions whistled down the wind—still there is so much in this extraordinary volume to stimulate reflection, and excite to inquiry, and provoke to earnest investigation, perhaps (to this or that reader) on a track hitherto untrodden, and across the virgin soil of untilled fields, fresh woods, and pastures new—that we may fairly defy the most hostile spirit, the most mistrustful and least sympathetic, to read it through without being glad of having done so, or having begun it, or even glanced at almost any one of its pages, to pass it away unread."—*New Monthly (London) Magazine.*

Illustrations of Universal Prog-

ress. A Series of Essays. By HERBERT SPENCER, Author of "The Principles of Psychology;" "Social Statics;" "Education." 1 vol., 12mo.

"The readers who have made the acquaintance of Mr. Herbert Spencer through his work on Education, and are interested in his views upon a larger range of subjects, will welcome this new volume of 'Essays.' Passing by the more scientific and philosophical speculations, we may call attention to a group of articles upon moral and political subjects, which are very pertinent to the present condition of affairs."—*Tribune.*

Thirty Poems.

BY WM. CULLEN BRYANT. 1 vol., 12mo.

"No English poet surpasses him in knowledge of nature, and but few are his equals. He is better than Cowper and Thomson in their special walks of poetry, and the equal of Wordsworth, that great high priest of nature."—*The World.*

www.ingramcontent.com/pod-product-compliance
Lightning Source LLC
LaVergne TN
LVHW011213110826
845150LV00006B/1424

* 9 7 8 1 4 2 5 5 1 6 9 1 8 *